positions asia critique

W0246569

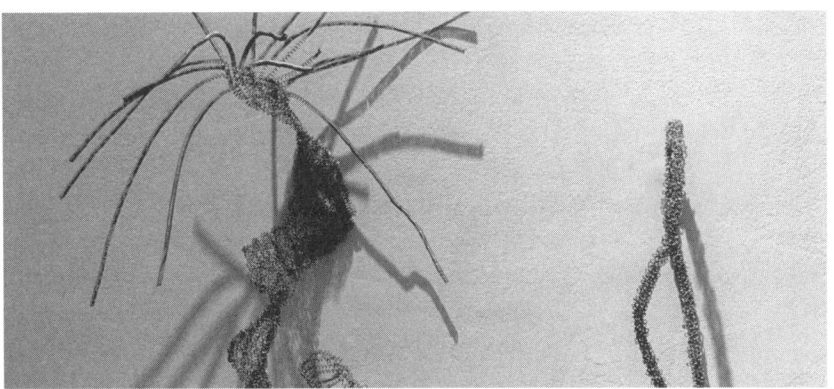

the end of area:

biopolitics, geopolitics, history

volume 27   number 1   february 2019

Contents

Guest Editors' Introduction

The End of Area

Gavin Walker and Naoki Sakai

A Genealogy of Area Studies

The term *area* may have a variety of connotations, but the "area" that is the central topic of this issue of *positions: asia critique* is thematically determined within the historical context of what is generally referred to as *area studies*. Broadly speaking, area studies designates a group of academic disciplines, initially conceived of during the advent of the Second World War and later institutionalized under the political climate of the Cold War in the higher education arena in the United States—at universities, research institutes, associations, and public foundations—whose raison d'être was originally to produce knowledge that would serve the global strategy and policy positions of the United States of America. Each of the disciplines under this heading takes an area as its legitimate object, and the identity of an area serves as the principle of disciplinary integrity and specialization. This concept of "area,"

positions 27:1   DOI 10.1215/10679847-7251793

therefore, used to constitute an object of inquiry in reference to which the disciplinary field measured its efficacy and improved its efficiency. In this protocol of disciplinary legitimation, area studies was not unusual or exceptional in comparison to other academic disciplines in terms of its internal epistemic constitution or its mode of self-legitimation.

Most academic disciplines in the natural and social sciences and the humanities organize and legitimate themselves around the specific objects of their inquiry. The configuration in the academic classification of disciplines is assumed to correspond on a one-to-one basis to the configuration of many aspects of human existence in the cosmology of the universe. The academic discipline of biology takes "life" as its object of inquiry; economics forms itself in reference to a putative object called "economy"; sociology legitimates itself by claiming to study the object of knowledge "society"; linguistics institutes its disciplinary protocols by competing over the most appropriate and efficient ways to study its own object "language." Each of these disciplines constitutes its own domain of specialization by forming an exclusive field of knowledge and by policing the boundaries of this field. And yet we are aware that each of these objects, without which an academic discipline can be neither legitimated nor constituted, is historically contingent. The very formation of these academic disciplines must be historicized: "life" meant something entirely different until the early nineteenth century; "economy" acquired a new epistemic and social function when it was completely alienated from the distribution of resources in a household and the gift exchange in social relations regulated by the order of kin, feudal, or monarchical lineages; "language" began to invoke a different domain of humanistic knowledge when signs and things in the world were clearly differentiated and related in a framework of representation.

In the sense that the object of an academic discipline forms and serves to legitimate knowledge produced in that discipline, area studies is no different from other disciplines in the social sciences and humanities. It too legitimates itself by claiming to study its own object—area—and specialize in it.

However, area studies is distinct from other disciplines in two particular ways. First, the object of inquiry is not one aspect or another of human existence, but rather an area and the group of people inhabiting it. In this respect, the knowledge of area studies is specific with respect to the geographic "location" of the object-people. An area renders as representable a certain surface of the earth where a certain "people" live. In other words, an area makes it possible to "locate" a place. Area studies is explicit, or one might say even candid, in its commitment to sovereignty: by appealing to the word *area*, area studies does not hesitate to display its primary relationship to its object of inquiry, the defining relationship between the knower and the known for this academic discipline by locating the object-people. Unlike sociology or linguistics, which is concerned with some abstract aspect of the world, area studies targets the population of a certain location as its object of knowledge. In other words, the primary relationship to its object is to know how it ought to be located and described, how its future trends and tendencies could be predicted, how to generate the most rational responses to its compositional elements: in short, to know the object of area studies is also at the same time to know how to *govern* that particular population.

Of course, our contention is not to insist on the exceptionally political character of area studies. Every discipline of knowledge production entails the practice of power in more ways than one. We would never deny that a particular formation of economics or linguistics either attempts to institute certain rationalities for social conduct or serves to install a new pattern of subject positions to know. However, what is significant about area studies is that, by appealing to the classification of humanity according to areas of putative belonging and inhabitation, it does not hesitate to exhibit an imperialist will to sovereignty, and it does not conceal the manifestation of colonial cartography according to which the surface of the earth is divided into multiple segments for remote control, whose sense of remoteness and distance will be further discussed in this introduction.

Each of these segments retains its unity, but the principle of the integrity of this unity does not derive from the needs and demands of the population inhabiting the area but rather from the strategic conditions of those who catalogue data and produce knowledge about it *at a distance*. This is why an "area" must not be confused with a "territory," as area studies has not been directly subject to the territorial state sovereignty. The integrity of this unity called "area," therefore, is constituted in accordance with the logistic needs and demands of a *remote* control. Accordingly, an area does not coincide with a national territory clearly demarcated by national borders even though, to a large extent, the disciplinary formation of area studies accommodates and even reinforces the basic laws of national sovereignty in the modern international world. For while national territory is a space marked for national sovereignty, area is not. The sort of sovereignty in correlation with which an area is constituted is not national in the way that the state sovereignty of the nation-state is; it is a sovereignty of remote control, perhaps called more appropriately the "global sovereignty" of the superstate.

Instantly, it is obvious that area studies is implicated in intimate but insidious relationships with disciplines modified by national or ethnic genitives, *Chinese* history, *Russian* literature, *Indian* philosophy, or *Japanese* sociology. Very often the content of knowledge is identical between national disciplines and area studies. Area specialists constantly consult the archives of national disciplines. This assessment invites us to the second feature that distinguishes area studies from other disciplines, national disciplines in particular.

One of the key and constitutive features of area studies can be found in its regime of *separation*, by which the knower is always and necessarily located *outside and at a distance from* the object of knowledge.[1] But it is important not to confuse this "outside" with territorial or geographic externality. Even if an area specialist lives, works and publishes inside the geographic space of the area, he or she is located in a temporality and epistemic regime that is separate and distant from the time of the indigenous. Johannes Fabian's famous formula is still pertinent here: "area" marks things and people for description, but these phenomena of description are deployed on the basis of the separation of two temporalities, of the time in which the known or the indigenous lives, and the time in which the knower or an area specialist narrates and positions him- or herself. That is, the "coevalness" of the knower and the known is denied (Fabian 1983: 31). The hierarchical organization of social relations, which Johannes Fabian sought to underline by his formulation concerning "the denial of coevalness" (32), persists and continues to determine how the knower positions him- or herself vis-à-vis the known in knowledge production in area studies. In essence, this separation is metaphysical in that it pertains to an area specialist's endeavor to be simultaneously both in the "area" and not in the "area," in a manner not entirely dissimilar to that of the missionary who is at the same time together with and separate from the local "natives" whom his or her mission is to proselytize.

It is in the sense of this disavowed "coevalness" that the knower is located *outside* the specialized area in knowledge production in area studies.

In contrast, national disciplines such as national history and national literature are built on an equally metaphysical presumption of *nationality*, a presumption of shared but empty temporality. To elucidate the metaphysics of the nation form, let us consult a classical definition of *nationality* offered by John Stuart Mill ([1861] 1972: 391): "A portion of mankind may be said to constitute a Nationality if they are united among themselves by common sympathies which do not exist between them and any others—which make them co-operate with each other more willingly than with other people, desire to be under the same government, and desire that it should be government by themselves or a portion of themselves exclusively." Nationality thus suggests a historically specific structure of communality, which refers to the gathering together of a "species" of people who share suffering as well as pleasure, the creation of a distinction between fellow countrymen and foreigners, the fostering of more cordial and stronger bonds with one's countrymen than with foreigners. It is living under the same government, enjoying self-rule, and disliking the idea of being subject to foreign rule; it involves independence and responsibility for the welfare of one's own country. With regard to the historically specific structure of communality, nationality means a modality of living together with one's compatriots, whose features are given in aesthetic terms. Therefore, Mill talked of "the feeling of nationality" and the nation as "a society of sympathy" (391–98). In terms of feeling and sociability, nationality thus defines insiders against the outsiders and delineates the spiritual space of communal interiority in which one's compatriots dwell. The communal interiority thus imagined is a spiritual space permeated by homogeneous and empty time: it is homogenous because it is nothing other than an abstract form of synchrony, not of simultaneity; it is empty because it is deprived of all the other concrete social qualities except for sharing.[2] Therefore, in national disciplines the knower and the known are not coeval either. This is precisely because the nation is an imagined community, in which each member of the nation is supposed to be copresent to other members in imagination, in the temporal modality of synchrony.

What nationality evokes is exactly opposite to what is implied by separation, which indicates the very absence of simultaneity and sharing, yet nationality does not indicate coevalness either, for nationality is a replacement for coevalness or a sort of *counterfeit* coevalness.[3] Precisely because both are alienated from coevalness, nationality forms a complementary symmetry with separation. And national disciplines are to foster this sense of spiritual communality without simultaneous participation, whereas area studies essentially belongs to the outsiders.

In the earliest phase of the development of area studies, it was portended that both North America and Western Europe would eventually be included in the list of areas for area studies (see Robert B. Hall et al. 1947). In its subsequent history, however, that prediction was never realized; neither Western Europe nor North America became areas around which the disciplinary fields of area studies were established. Interestingly enough, nominally European though its designation suggests, Eastern Europe was listed as an area, and we have seen the development of East European or Slavic studies as an area studies. What prevented Western Europe and North America from being postulated as areas are the power dynamics we refer to as "the discourse of 'the West and the Rest'" (Hall 1996: 189). What is at stake for area specialists in determining whether or not a certain region and the people within it are designated as an area is the question of *positionality*. To thematically postulate a region and its people as an area is to nonthematically postulate the position of area specialists as the West. In other words, area specialists could not discard their desire to fashion themselves so as to occupy the position of "the Westerners" and so continued to identify themselves in the *schematism of cofiguration* (see Sakai 1997). At the end of the day, it will be revealed that what constitutes area studies is precisely the identity politics of the West.

Consequently, the following implications can be deduced from the fact that a certain population and the geographic region it inhabits are postulated as an area whereas a certain kind of humanity and their residential regions are never called "areas." The first condition proceeds thus: two dimensions of the disciplinary identity are available for area studies, on the one hand, in terms of a particular aspect of the cosmology of the world, and, on the other hand, in terms of the provincial name of the place and its population. Thanks to this arrangement, area studies is formed as an *interdisciplinary* discourse. An area specialist may well be a sociologist or literary historian according to the first dimension of classification, but he or she can be a specialist of South Asia, for instance, according to the second. From the outset, therefore, area studies was conceived of as an interdisciplinary formation precisely because of this doubled or two-tiered constitution of the discipline.

The second condition dictates that since such a postulation of an area is possible due to the working of the discourse of the West and the Rest, an area thus determined must be located on the side of the Rest, as an other to the West. Precisely because the knower and the known must be *separated* from one another and *located* in the West and the Rest respectively, North America and Western Europe could never be included in the list of areas. This is to say that the separation thus instituted is of a civilizational sort, between one civilization and another, and of an anthropological difference, between one type of humanity and another (see Solomon 2014). This explains why, in American academia, those in area studies quite often demonstrate intense antipathy and resistance to the potential merger of area studies with ethnic studies. Many of the area specialists specializing in East Asia, for example, do not hesitate to publicly announce their refusal to work together with specialists of Asian American Studies; they often insist on geographic distance, a gap in the degree of modernization, or civilizational difference that separates areas of East Asia from the North American continent.

When it comes to the question of their own positionality, area specialists not infrequently invoke—or tacitly assume—exoticist formulations such as "the East and the West never meet" or "East Asian heritage remains entirely heteronomous to American or European experience," and so on. Do they want to say that the vast distance of the Pacific Ocean keeps the West separate from the East forever, whereas the equally vast distance of the Atlantic Ocean holds the West together on both shores of the Northern Atlantic? It goes without saying that what lies behind all these inconsistencies is the identity politics of the West.

Despite a constant appeal to various tropes of distance and separation, however, it is absolutely crucial to keep in mind that, in the final analysis, neither the West nor the Rest is a geographic entity; nor is it a substance that is cartographically identifiable. Therefore, it is not surprising that African Americans, for instance, are scarcely recognized as "Westerners" even though, for generations, they have received the same education—even within a segregated system—spoken the same language—even though clearly marked by dialogical difference—and believed in the same religion—even though they scarcely shared the same church space—with their family lineages going back much further in North American history than those of many white people. Supposedly North America is located in the West, but it suddenly gets disqualified as a geographic part of the West when the topic is African Americans, who undoubtedly have lived for many generations in the United States, a space recognized—largely since World War II—as part of the West or even as the center of the West.

Obviously the West and the Rest are always provisional markers depending on many contingent circumstances—involving a wide range of variables such as social class, race, cultural capital, religious affiliation, industrial wealth, and so forth—and the very dichotomy of the West and the Rest does not make any coherent sense unless we take into account the microphysics of the power relation operating in specific locales in which the West and the Rest are indicators of certain positionalities. Neither the West nor the Rest is, as a matter of fact, a geographic category, yet people act as if these civilizational identities were geographically determinable by ascribing them to points on the map of the world. It is in this posturing toward cartographic presentation in which places are determined as locations that the civilizational identities become geopolitically effectual. For this precise reason, we require—theoretically and politically—*the dislocation of the West (and the Rest)*. It is also for this reason that we cannot conflate an area with a territory, or area studies with national disciplines.

The dichotomy of the West and the Rest serves to distribute positionalities or different strategic positions among which certain conducts are routinized. The positionality of the knower, who fashions him- or herself as a Westerner, is constituted in a particular situation of interaction with the known. Even though the positionality of the knower is fantasized as separate and distanced from the positionality of the indigenous, it is performatively actualized simultaneously with the postulation of the known. The knower is identified as such only when the opposite positionality, the known, is postulated. Even though the exercise of control is somewhat imagined to be remote and indirect, the operation of the discourse of the West and the Rest serves to effectuate the simultaneous actualization of the colonial configuration of strategic positions. However remote and indirect the known may appear in the gaze of the knower, the microphysics of the power relation works in the very institutionalized interaction of the knower and the known.

This is one of the reasons why we find the defining feature of area studies in its governmentality, in its will to know how to govern the population of an area. Yet knowledge production in area studies is not exclusively organized with a view to repress the initiatives of the indigenous or prevent them from having their identity. We do not simply believe that area studies must be denounced because it is repressive. What must be emphasized is rather the *productivity* of area studies, thanks to which the known are given their identities, encouraged to occupy the positionalities ascribed to the Rest—Asia, Africa, Oceania, or the Americas. It is precisely in this context that the question of area must be understood in conjunction with that of biopower. To govern the population is not merely to impose the knower's purposes on the known, force the known to act as the knower wishes, or oppress the will and aspiration of the known. On the contrary, what is at issue is the governmentality of the area—let us note that it is not the governmentality *over* the area—and this governmentality is actualized in the very postulation of the area; it can be discerned in the reaction of the known, in their resistance to the knower's gaze, in their endeavor to produce knowledge in their national disciplines in reaction to knowledge produced in area studies. The population of the area is objectified by the knower in area studies, postulated as a unified target of description, measurement, and assessment, and envisaged as a self-transforming agent whose collective behavior it is the task of area studies to decipher, interpret, and predict. Nevertheless, the governmentality of the area cannot be exhausted in the catalog of obligations that area specialists are expected to fulfill, for it is not a one-way operation exercised by the knower on the known. More importantly, area studies can affect and mobilize the population of an area through the production of knowledge about this population itself. In short, the governmentality *of* the area that can be discerned in area studies is *productive* rather than *repressive*, from the viewpoint of the known or the indigenous.

Elsewhere we have discussed the Orientalist mode of knowledge production in area studies and how it serves as a subjective technology by which the knower, not the known, constitutes him- or herself as a subject (see Sakai 1997: 24–25).[4] It is widely acknowledged that, as Edward Said amply demonstrated in his epoch-making publication *Orientalism* (1978), the Orient is postulated and generalized as an object of knowledge in the discourse of Orientalism and that, in this discourse, the Orient is expected to fulfill a number of requirements: its subordination to the Occident in terms of civilization and scientific rationality (the Orient is less civilized and incapable of expressing itself rationally); the Orient lacks rationality and therefore is mute, as far as its self-expression in rational language is concerned, so that it requires the Occident to speak on its behalf; its passive positioning vis-à-vis the gaze of the Occident (the Orient is exotic/mysterious and "feminine" for its passivity); a belated status in evolutionary teleology (for the reason of which the Orient is always bound to follow the Occident in progress); and so on. Undoubtedly these requirements, which the Orient is supposed to fulfill in its identification as the Orient, *are inverted forms of desire for the identity of the Occident*. Unless the Orient is postulated as such, the Occident cannot identify itself. In this respect, Orientalism is a subjective technology for the Occident; Orientalism is, in fact, Occidentalism turned on its head. In our critique of Orientalism in the past, therefore, the focus has been predominantly on the identity politics of the Occident. In this issue of *positions: asia critique* on the question of area and area studies, however, we are less concerned with the identification with the West on the part of the knower, those engaged in the production of knowledge in area studies. What we want to undertake is to examine how the known, not the knower, are also affected in Orientalism, how they react to and resist the imposed image of area by area studies, and how they voluntarily consolidate themselves as the population and thereby get incarcerated in the discourse of the West and the Rest. In short, in the rest of this introduction we want to pursue the way in which the known are implicated in the working of power as a result of the putative separation of the West and the Rest, how the indigenous form themselves as an integrated population in the gaze of the knower, and how the identity politics of the indigenous is complicit with the power of area studies' will to knowledge. Of course, we are talking about what we have referred to elsewhere as *civilizational transference*.

The International World and the Schema of Civilizational Difference

Let us make no mistake. Our belonging to a place is neither merely factual nor existent on a descriptive basis; it has never been natural in the sense of a place just being there for us to belong to without our intervention and production. In belonging to a place, the place is constituted as a location in the very act of belonging. In other words, our location in a place is always of our own making-doing, of our *poiesis*.

Some readers might call into question the use of the genitive pronoun "our" here, its appropriateness and conceptual underpinnings in the context of a discussion of area. Can such an "our" or "us" be easily posited in the systematic distribution of positions characteristic of the apparatus of area that we have been investigating? Without delving deeply into this question, in which the problem of area would also be linked to the concept of the *common*, the genitive can be replaced by "one's" for the moment. Thus we say here: "one's belonging" instead of "our belonging."

One belongs to a house, a town, a province, a country, and an area. All these places one might belong to are supposedly on the surface of the earth, existing as marked segments of land with locations determined in a configuration with other land markers and land posts. It is most likely that all of these places of belonging can be identified in terms of circumscribed territories on globally extended maps, whose scale may vary from the size of a land register—in the case of a house—to the size of a terrestrial globe—in the case of a country or an area (in the sense of area studies). Let us not forget that these places must be, above all else, on maps in order for them to be registered as locations determined in the system of coordinates. They are neither reflections nor reproductions of these shapes and relations that are found in themselves in nature; their determination cannot be subsumed under the principle of *tracing*, the goal of which is to describe the de facto state. Only through an act of *mapping* rather than *tracing* can these places emerge determined as locations on certain kinds of maps—let us not forget there are many different kinds of maps; in other words, there are many different ways to establish one's relationship with a place.[5] An area does not escape this general rule of territorialization, and it is in the formation of an area that the very mechanism of territorialization is visible. So, to inquire into the formation of an area is to pay attention to the act of mapping through which a place of belonging in general comes into being, on the one hand, and the process of bordering by which a place is circumscribed and reintegrated into the order of the binary logic of global territorialization, on the other.

This is precisely where we see the significance of the concept of nomos and the importance of Carl Schmitt's inquiry into the appropriation and expropriation of the territorial surface of the earth for our study of area and area studies.[6] As Schmitt (2006) outlined in his *The* Nomos *of the Earth in the International Law of the* Jus Publicum Europæum, the general rules of territorialization in the system of international law are the rules of what is understood as the *international world* in the last four centuries or so.

The international world is a name for the schema—image, figure, or plan—of a global geopolitics whose projection is determined by two constituent principles. Above all else, it is important to keep in mind that the international world is in the order of image and figure, but this by no means implies that this image or figure of the world is merely descriptive; rather it is prescriptive in the sense that it projects and institutes things according to predetermined directives. Together with the system of international law, the image of the international world has served as an assemblage of commands, norms, and standards, that is, as a regime. Hence, we do not argue that the international world came into being in the seventeenth century; instead we propose that the geopolitics of the world began to be transformed toward the ideal of the world projected by the schema of the international world that prevailed at the time. By the twentieth century, however, the international world was virtually everywhere on the territorial surface of the earth because a sovereign state that was not legitimated by international law could scarcely be viable by then.

The system of sovereign nation-states was first developed with the Treaty of Westphalia (1648) during the initial period of the imperial-colonial era when the world was divided into two realms, one governed by international law and the other subject to the discretion of colonial powers. The first realm was then called *Europe*, which came into being as an *international* space. As the system of international law expanded to cover the entire globe,

Europe was also called *the West* while the second realm was equated to *the Rest* of the world.

The first of the two constituent principles is the spatial externality of one state sovereignty to another. It is important to note that the new modality of sovereignty, which initially appeared in the imperial-colonial era (sixteenth to nineteenth century), did not concern the national space; the national space was structured by the markers of distinctions such as class and rank rather than those of nation and ethnicity, but state sovereignty concerned first and foremost the international space of a global system. What first emerged in Europe in accordance with the *Jus Publicum Europaeum* (Eurocentric Public Law) was the modern state formation characterized as territorial state sovereignty. It is only in the eighteenth century that a few exceptional state sovereignties—namely the newly independent United States of America and the Republic of France—acquired a new form of legitimacy called "the nation," and thereby justified themselves in terms of the specific form of the territorial national state sovereignty. Today the overwhelming majority of existent states legitimate themselves according to the model of sovereignty of the territorial national state, but the most crucial transformation brought about by the institutionalization of international law was the state's relationship to other states, namely, the *inter*national relation. In this precise sense of an international space, the modern state became dominant.

By definition, the international space for international law was one of commensurability, a space in which different states were compared and mutually recognized. Only when states are juxtaposed and compared to one another in the space of commensurability is the state's relationship to other states regulated and streamlined by the system of international law. Therefore, the first characteristic of the international world that we want to draw attention to is a space in which comparison is possible, the space of commensurability in which all the units of sovereignty are "countable" and "comparable"; the internationality of the international world is thus built on a comparative operation by which one state's sovereignty is recognized by other sovereignties, and one state's territory is marked and distinguished from the territories of other states. In this sense, it is not only that all units of sovereignty exist in a relation of countability between each other; the international world is a regime in which the territorial national state and its particular form of sovereignty also comes to be internally measured, insofar as it "counts for one."

From the spatial externality of one state sovereignty to another it follows that this space of commensurability is supposed to be regulated by a binary logic by which an inside is unambiguously distinguished from an outside. The binary logic is an organizational principle for *tracing* rather than for *mapping*.[7] But was the generative rule for the international world nothing but the operation of mapping? What is suggested here is that the very formation of the international world through mapping is repressed and deliberately overlooked in the international world because it is this mapping that discloses its historicity and repudiates its fantasy of eternality. It is as if the divisions and segmentations of the international world were simply reflective of a naturally given regime, as if the order of internationality were merely descriptive. Area studies have legitimated themselves on this presumption or pretension of descriptive factuality or positivistic nonreflectivity. Our investigation of area and area studies will focus on this deliberate oversight, which has allowed an area to be misrecognized as a territory, and because of which an area has served as a mediation for anthropological difference and civilizational transference.

A space of commensurability is necessary for the supposed integrity of territorial state sovereignty, because the unit of counting cannot be constituted unless its inside is clearly marked from its outside. The internal cohesion of state sovereignty is brought about by a sort of *enclosure*—a classical Marxian term associated with what is generally called "primitive accumulation"—an operation of circumscribing a territory in an international space. What must be emphasized here is that an enclosure also was and is a comparative operation, and there must be a space of commensurability for comparison to be possible (see Walker 2011).

The interior of the sovereign territory is thus distinguished from its exterior, and the site of distinction is instituted as the national border (see Sakai 1991). Perhaps for the first time in history, the presumption was authorized that the juridical authority of the sovereign is to cover the entirety of an enclosed territory constituting a homogeneous space, and also that one place must be subjected exclusively to one state sovereignty. The system of international law does not tolerate any appropriation of a place to multiple states.

Prior to the introduction of the international world, there used to be many places belonging to multiple state sovereignties, but these ambiguous places were eliminated one by one. The best example of this can be found in the case of the Diaoyu/Senkaku islands that are disputed between the People's Republic of China and Japan. Taking the concept of national territory for granted, both sides claim that this group of small islands in the East China Sea must have belonged exclusively to one state sovereignty or another. Neither of them can afford to acknowledge the historical truism that just over one and a half centuries ago there hardly existed any consensus shared by the Qing Dynasty and the Tokugawa Shogunate about these islands belonging exclusively to one state sovereignty. The concept of national territory itself did not exist then. Clearly Okinawa or the Ryukyu Kingdom, which offered tribute to both the Qing Empire and the feudal domain of Satsuma, was another case. Many other similar cases in Western Europe and Southeast Asia are known.

It is not difficult to see a similarity between the integrity of national territory and the unity of a language. As we have elsewhere theoretically demonstrated in relation to this historical change, this is why the introduction of the international world carries an affinity with the transformation of the regime of translation (Sakai 1991). Besides territory, the question of language is indispensable for an understanding of internationality. Is language a unity that is "countable" and that can be juxtaposed with other unities of language? Here, we face the exact same question concerning the space of commensurability. The crucial point is the process of *bordering* in which a space of commensurability is constituted where one unity is compared to another. Of course, this process is generally called "translation" when the topic is language,[8] but what is of decisive importance is that there are some regimes of translation in which language is not necessarily constituted as a unity, in which one national or ethnic language is not distinguished from another national or ethnic language and thereby the nationality or ethnicity of a language remains ambiguous or indeterminate.

The second of the two constituent principles is the very reason why the system of international law was called the *Jus Publicum Europæum* from the outset. The international space in which the juxtaposition of territorial state sovereignties was possible was marked as *Europe* or *the West*—as opposed to the *Rest* of the world.[9] Historically, until the twentieth century, the international space of commensurability did not immediately cover the entire territorial surface of the earth. On the contrary, it could be marked as such—initially as Europe and later as the West—only in contrast to excessive spaces in which the order of internationality was not valid. Yet the international space of commensurability assumed and anticipated the presence of these excessive spaces, and only based on the assumption of the anticipated subjugation of excessive spaces to the sovereign state of the international space of commensurability did it constitute itself as Europe.[10] This particular modality of Europe's presence in relation to the rest of the world was later called *Eurocentricity*, and, as we know, Europe's dependence on the presence of the excessive spaces—the Rest—has been conceptualized as "modern colonialism" or the colonialism of the international world.

Consequently, we must discern two moments in the Eurocentricity of the *Jus Publicum Europæum*. First, the binary of Europe and the rest of the world is supposedly of geographic positionality, of geopolitical location projected onto the map of the globe. These two kinds of locations, Europe and the Rest, are assumed to be in the geographic order and are accordingly independent of each other. But they are not merely geographic locations. They are mutually determined. It is absolutely imperative not to overlook the mutual implication of Europe and the Rest. Second, therefore, Europe and the rest of the world are cofigured, implicated in each other, so that Europe cannot be identified as such without reference to the rest of the world. It goes without saying that, as the term *rest* clearly indicates, the rest of the world presumes and constitutes itself after the presence of Europe. It is in this sense of mutual implication and mutual constitution that Europe (the West) and the Rest are projected through the schematism of cofiguration. In other words, it is what Gilles Deleuze and Félix Guattari (1987: 424–73) called "capture" on a global scale.[11]

By referring to "capture," we are trying to underline that neither the West nor the Rest (of the world) can be talked about without reference to one another. This binary serves to project both identities—of Europe, or the West, and the Rest—by integrating a variety of social relations and locations into the global order of modernity by differentiating and contrasting them. In continually inserting a difference of civilizations—colonial difference or anthropological difference—this regime integrates and structures the world. Not only by distinguishing Europe from the rest of the world but also by integrating them and projecting the potential appropriation of Europe in the Rest, this regime serves to structure the entire world around Eurocentricity. From this regime of cofiguration follow the accompanying notions of the West-as-a-normative-value and of modernity-as-an-unfinished-project. By explicitly calling into question the binary of the West and the Rest, therefore, we are advocating neither the rise and decline of the West nor its universalization or provincialization; nor does our perspective amount to the disowning of heritage from the past such as the rejection of putatively "Western knowledge" for the sake of local tradition. It would be very obvious that the West cannot be referred to even in the trope of an organic unity that grows or languishes. What is at stake in our inquiry into area and area studies is to dislodge both the West and the Rest from the identity politics of civilizational transference and from the logic of anthropological difference.

The modern era might be characterized as the time of the *Jus Publicum Europæum* in which this structure of Eurocentricity has been implemented, to the extent that it has been taken for granted not only in Europe but also all over the world. Since the aftermath of the Second World War, however, it is increasingly difficult to justify the Eurocentric posture of the international world. Let us not forget that, paradoxically, the internal decay of the Eurocentric system of international law was brought about by the very success of this system itself. In order to reject colonial domination, many peoples and regions began to adopt the basic premises of territorial national state sovereignty. Anticolonial independence movements, almost without exception, aspired to achieve the form of nation-states as their objectives. But through the extension of the *comity*—the formal legal reciprocity—of nations across the face of the globe, the supplement of exteriority known as *civilizational difference* has reached a point of saturation, which is also a point of crisis. Civilizational difference indicated, in its implementation in colonial governmentality, the economy of spatialized lawlessness that defined the West by separating its competitive rule of law from the Rest that was available for lawless, infinite violence. Although this lawless violence is still exercised by the United States and a few Western European countries such as Britain and France, it is increasingly difficult to justify this use of colonial violence in many parts of the world.

The use of drones in such areas as Afghanistan best symbolizes the economy of spatialized lawlessness that used to separate the West from the rest of the world. Afghanistan is a reserved area of the Rest available for precisely this sort of extralegal violence without end. Unambiguously it shows that the *Pax Americana* marks not an end of colonialism but an extension of the schema of the international world, the continuation of which requires the maintenance of an economy of spatialized lawlessness. But this new "cramped space" in which the extension of the international world's spatial schematic currently operates also involves (as always) the politics of knowledge.

In 2006, retired Major General Robert H. Scales wrote a text called "Clausewitz and World War IV" in the American *Armed Forces Journal*. This text, openly advocating a new orientation of military capabilities away from the direct stockpiling of munitions and toward a form of stockpiling of cultural knowledge, ends on the following note:

> We are in for decades of psycho-social warfare. We must begin now to harness the potential of the social sciences in a manner not dissimilar to the Manhattan Project or the Apollo Project. Perhaps we will need to assemble an A team and build social science institutions similar to Los Alamos or the Kennedy Space Center. Such a transformational change is beyond the resources of a single service, particularly the ground services.
>
> Thus a human and biological revolution will have to be managed and driven by the highest authorities in the nation. (Scales 2006)

This "harnessing" of the social sciences reached its zenith in the form of the US military's program known as the Human Terrain System (HTS), a program devoted to the development of "a social science based research and analysis capability to support operationally relevant decision-making, to develop a knowledge base, and to enable sociocultural understanding across the operational environment" (quoted in Patterson 2016; see also Bartholf 2011). HTS expresses the new stage, the new modality of the operation of area studies, which has transitioned from an entirely practical, logistical orientation to a new ideational, strategic formation. By harnessing into one productive space the reproduction of imperialism and the production of knowledge on the basis of area as a gradient of investment, HTS expresses in its formation the current stage of our politics of knowledge. In this sense, HTS—although it has now been discontinued as a project—exemplified precisely the transition from the traditional "strong" area studies, in which specialists would produce knowledge entirely within the university, and this knowledge would then circulate into policy circles, to a new mode in which the internal division between knowledge and application has been short-circuited. Today, the military and state no longer need to *indirectly* intrude into the production of knowledge by actively intervening in the university itself, because, instead, these instances of the hegemonic social relations now employ knowledge workers *directly* in the maintenance of imperialism. Just as the media was "embedded" with the Americans, so too now "area specialists" are literally members of the military. This situation, in which the old formation of area studies has been both superseded and perfected, requires us to think clearly about what possibilities and limits lie within the logic of "area" itself on which such knowledge is produced. HTS and its correlates beautifully illustrate how the superstate reterritorializes the schematic of the world, the cartography of civilizational difference in a new improvised way. It is not the case that the deterritorialization of capital has created an "untethered" world without borders—everything that is deterritorialized is also reterritorialized. Rather, the ongoing reterritorialization of the world schema of civilization is multiplying its instances of border drawing; it is an improvisational flow, the superstate's overtone of mapping.

While this appears to be a scenario in which the multiplication of enclosures and confinements produces a dense and inescapable space, it is the very multiplication of these bordering instances that thereby produces a multiplicity of open and living productions of subjectivity. In thereby increasing our encounters and suspensions with respect to the reproduction of the social field as a bordered one, we are also constantly encountering *the bordering function as such*, that is, we are increasingly exposed to the creative-formative dimension of the existing order's ongoing maintenance. This does not mean that politics has returned to the stage, so to speak, but rather exactly the opposite: as Paul Virilio has long pointed out, "The globalization of liberalism is a deterrence of politics" (Virilio and Lotringer 2002: 59).

Broadly speaking, therefore, it is hard to identify cultural and political markers by which the West can be *unambiguously* distinguished from the Rest today. Elements associated with "Western modernity" can now be found in places that have conventionally been excluded from the West—often in forms that are more "authentic" than those found in the West itself. Is the life style of upper-middle-class residents of Seoul less "Western" than that of the poor whites of Appalachia or in the southern states of the United States by virtue of the fact that they live in the non-West? Is the social infrastructure found in Bulgaria better adjusted to the pursuits of modern rationality than that found in Shanghai by virtue of the fact that Bulgaria is supposed to be in "Europe"? At the moment when the global expansion of the two universal forms of capitalism—the commodity and the nation-state—is finally accomplished, historically for the first time such civilizational distinctions as the West vs. the East and Europe vs. Asia appear as what they essentially are: void of any specific content and thus absolutely ideological (see Walker 2012a). Therefore, no longer are there any grounds whatsoever to substantiate the distinction between the West and the Rest. Or, put another way, it is no longer possible to continue to disavow that the West is floating and dispersing (with the tides of domination); but it is equally important to note that the West is *not* declining. Hence, our inquiry must be characterized as the dislocation of the West.

The End of Area and the Biopolitics of Investment

Under such historical circumstances, how do we assess the future of area and area studies? What has happened to the notion of area and to the disciplines of area studies? Allow us to present some preliminary answers first: what we mean is precisely the end of *area*, but not necessarily the end of its disciplinary expression, *area studies*. Area studies has not ended, but it instead becomes increasingly tied to a schema of the world that no longer exists (see Walker 2019, this issue). What is in question for us is not "the end of area" as in the end of the importance of specific knowledge, linguistic study, or historically particular circumstances; the end of area means just the opposite, the end of the *schema* area, the end of the *regime* area, the end of this epistemic poietic device through which knowledge is "nationalized" and thereby rendered "inherent" and "natural." It is this ending—and inherently, there-fore, a new and possible *beginning* of another knowledge, another politics, another history—that animates this issue of *positions: asia critique*.

Unlike the situation of the 1990s, in which we frequently heard that the expansion of world capitalism and its accompanying neoliberal political consensus would be the guarantor of a new and open future, today almost nobody bothers to make such claims except the most determined right-wing ideologists. We are living rather through a moment in which the supposed closure of history has reopened. It is no accident that Francis Fukuyama (2012), who once famously declared the "end of history," has suddenly now referred to our time as "the future of history," calling into question the once-cherished "permanency" of capitalism and the international world—the world as constructed from the plurality of nation-states that are geographi-cally external to one another. After all, it is now fully clear that political integration, the integration of national capitals, and the development of superstate forms of political direction have resulted in no substantive resolu-tions of "the national question." In other words, whereas we used to hear that globalization was merely the name for an ongoing resolution of conflict on the basis of the nation-state, today we are exposed to the precise inverse of this logic: instead, globalization discloses that it is the raising of the national question to a set of political techniques not only for the governing of the domestic situation but for world capitalism itself. That is, our era is not one in which the political technology called "area" has been superseded in favor of a more open, more global form of governance—rather, the political force of "area," which was previously one important principle of the system of the international world, has now been raised to *the* principle of the suppos-edly "integrated" world, its highest logical operating point. In an important work, Federico Rahola points out that rather than the "age of access" pre-dicted by Jeremy Rifkin and others who theorized a glorious eclipse of the industrial sequence of history in the 1990s, we are living through a historical present which might better be described as an "age of excess" (Rahola 2003: 34), not in the sense of overdevelopment or something excessive, but rather in the sense that the techniques and modalities of the political governance of capital are increasingly concerned with humanity's excess over its previously established boundaries, orderings, and zones of territorialization, excess over the classification sustained by the system of the international world in which the territorial unity of the state is posited external to those of other states. When Rahola refers to our moment as the "age of excess," he is also pointing to the Rest, the remainders or remnants, the leftovers. As we have suggested above, the Rest is no longer geographically located outside the West. The Rest is everywhere and inside the West as well.

In other words, what matters increasingly for capital and the nation-state (and the volatile amalgamation between them) is not "access," the opening and freeing of the previous limitations and borders. Rather, capital and the nation-state derive a dynamism of operation from the *excess* that is generated by the enclosure of new ideational spaces, new temporal sequences, new sites of subjectivation, and the erecting of borders where previously there were none. It is this set of techniques, this *cartography* of capital and the nation-state, that furnishes the ground of the continuity of "area" as a subjective technology. Area is a quintessentially modern form of the classification and ordering of the world, and it is inseparable from the advent of the modern state. Prior to the beginnings of the modern state forms of organization of territory and population, we cannot speak about the sense of "area" that we mean. "Area" as a form of classification is not something primal or originary, it does not precede the state—it is important to distinguish area as an epistemological-political technique from the common-sense or everyday understanding of area as simple and self-evident spatial unit of measurement that is bounded and gathered improvisationally in terms of certain given social circumstances. Rather, "area" is a primordial move to capture people and places into the order of the regime of cofiguration, a synthetic operation by which to unify a population and a geographic region by differentiating according to the logic of colonial difference. Thereby, we want to give a specific social and historical meaning to "area," that is, "area" as it appears in the phrase "area studies."

But what is this specific sense of area? We intend by this formulation to indicate the undercurrent or substratum of the territorial grounding of the population through techniques of unification and semiotic effects such as "culture," "language," "ethnicity," and so forth that are demanded by the system of the international world. In this respect, area cannot be confused with the territory of the territorial national state sovereignty. For specific historical reasons, China, Korea, and Japan are conventionally recognized as areas and national territories, but apparently such areas as Southeast Asia and Latin America cannot be registered in the order of the international law.

In this sense, we want to develop a discussion of "area" as an essential operation for the internal cohesion of the governing capacity of the state in parallel to the question of "population," a form of the investment of state power within life, what can be called, after Foucault, *biopower*. Yet, it is important to stress the remote position of the state in relation to the population it aims to govern. Perhaps the trope of the drone captures the essence of an area as it operates in area studies. Presumably the population in an area is always *separated* from the original source of the government just as the site of violence is distanced from the position of the pilot who maneuvers the drone and triggers violence. Therefore, an area constitutes neither a "nation" as a particular form of population nor a "territory" as a location for the territorial state sovereignty. Instead, an area always belongs to somebody else's "nation." For the practitioners of area studies an area does not mark their own belonging to a community or to a place. One cannot overlook this constitutive moment of *separation* between an area and a nation, between an area and a territory, for the operation of biopower.

On the other hand, in the disciplines of the humanities since the outset of modern universities in the eighteenth century, which we might as well summarily call the knowledge system of national translation (see Sakai and Solomon 2006), these parallel operations of articulation of "territory" and "population" are required by the state in order to give to itself an image of community called "nation," an image that folds back into itself in order to naturalize the modern form of belonging to the nation-state and create a heuristic measuring device for "normal" and "exceptional," that is, "majoritiarian" and "minoritairan," positionalities within it. But this arrangement of internationality essentially dictated by the first constituent rule mentioned above, an arrangement of configuration of one nationality external to other nationalities, cannot be extended to explain the features of an area. An area is a typical manifestation of the second constituent rule in which the focal point is not the identification of population as a nation but rather a differential integration of populations into the order of colonial difference.

Positions and identifications such as "the West" or "Asia" are not, in essence, concrete entities; rather they are increasingly independent of geography and based in forms of "cultural capital" or certain relations of positionalities that are "performatively presented." These supposedly regional identifications are a series of effects derived from the ways in which people invest in the acquisition of such qualifications. But the problem is, why do these temporary identificatory zones or grids of locatable identification always need to be retrospectively territorialized? Why does the logic of the specificity and explanatory power of "area" continuously renew itself? Why are the bordering effects of the modern forms of the internationality (increasingly identified with the "global police" function) reinforced precisely at the moment when the border itself becomes historically exceeded? What connections and lines of inquiry can be drawn between the bordering effects on the level of epistemology and the policing of the border so essential to the operation of contemporary biopower? This global biopower—an increase of disciplinary power that overflows the frontiers of the living—is a method of governing the foreign population and of controlling the biopolitical elements of human activity at a distance. This biopower grounds itself in the mechanism of area as a means to order, combine, separate, and classify life at a distance.

We know today that the form of area studies is increasingly irrelevant to the contemporary social reality we are living through, that it relies on an image of the world that is increasingly outdated; yet we nevertheless have to acknowledge the lasting power of "area" as a form of investment in knowledge production. Area has never been a substance but rather an intermediary zone of grounding between "*anthropos* (as against *humanitas*)" and "territoriality."[12] What sustains the practice of area studies is the imaginary relationship among projected positionalities in terms of which the identities of the observer and the observed are figured out. In other words, area is a technology according to which elements—which may or may not have been thoroughly heterogeneous to each other—are gathered and redeployed as a point of reference for a variety of social—racial, class, religious, gender, and so forth—distinctions. More generally, however, area is the logic by which biopower is articulated to the geopolitical ordering mechanisms according to which the world image serves as a framework of cognition in the lived world.

This is why, for instance, we cannot escape the logic of area merely by identifying its fictive character—it is precisely *because of* its fictive nature that "area" has been a lasting force in the operation of identification. Area in this sense is the grounding movement itself, the most recent stage in the deployment of the society of control theorized by Foucault: it is a "state of government that is no longer essentially defined by its territoriality, by the surface occupied, but by a mass: the mass of the population, with its volume, its density, and, for sure, the territory on which it is deployed, but which is, in a way, only one of its components" (Foucault 2004: 113; 2007: 110; translation modified). In general, he identifies the transition in mechanisms of discipline to those of control in their origin from "pastoral power"—this shift can be understood too as a shift toward the management of something wholly different: the population.

Population is a name for a sequence of strategic deployments that are concatenated through a process of articulation between state territoriality, the socioeconomic attempt to fix the role and supply of labor, and the process of the formation of national language as regulative regime. Marx, for instance, locates this total operation of articulation within what the tradition of English political economy referred to as the process of *primitive accumulation*, an entire field of problematics intimately related to the formation and maintenance of the territorial state. We should not underestimate Marx's location of the origins of capitalism in the process of primitive accumulation's establishment of racial hierarchy—Foucault emphasizes exactly this same point in the development and deployment of biopower, that is, he repeatedly shows how racism became inscribed as the basic mechanism of power in the form of the modern state system and internationality. In fact, the form of national sovereignty itself can scarcely function without becoming involved in and complicit with racism (see Walker 2012b: 122–147; Walker 2016, esp. chap. 1).

Today, many of the most incisive critiques of contemporary biopower and its accompanying geopolitics tend to misread or ignore the theoretical problem of how our current moment requires us to consider not only the persistent inequalities of the world order but also *how* this order or civilizational difference was itself formed, as well as *how* and *why* it is maintained and managed. Too often, responses to contemporary *globalization* refuse the logic of imperial nationalism in the name of another state project (whether European, pan-Asian, etc.) and end up reproducing the rhetoric of civilizational or anthropological difference—but we argue that only through a properly biopolitical response to contemporary biopower can we create new forms of encounter, new engagements in sociality that do not rely on the process of identification constantly renewed by the logic of area. Precisely because old civilizational differentiations are totally devoid of empirical bases are urges to restitute the West and the Rest all the more intense. At the historical juncture where the future of area studies is most uncertain, the history of area studies in North American and European universities must be revisited from this perspective. Moreover, the formation of national disciplines—national literature, national history, and so forth—in Asian universities must be reviewed also from this perspective. We are concerned with the initial formation of area studies in the United States, but we are overlooking neither its precedent histories—how US area studies was reconstituted out of the European studies of colonies and geopolitics and Japanese colonial scholarship—nor its consequent histories—how nativist knowledge was formed in reaction to the area studies about the local tradition. We want to inquire into not only why area studies was able to actualize US imperialist policies but also how it laid the cornerstones for the nationalisms that are complicit under the Pax Americana.

In other words, it is not enough to identify and criticize the "tendency" aspect of area studies: many knowledge workers are today aware that area studies is tendentially becoming irrelevant in the world, superseded by the emergence of new relations and forms of relation in the nexus of socially available knowledge, density, and concentration of technical factors and the emerging superstate level of sociality. As we frequently hear, the expansion of global capital and its concomitant supporting forms of social relation have superseded many of the old boundaries, dislocating the identification of wealth and territoriality that characterized the moments of the imperialist stage and its immediate aftermath. The logic continuously put forward within the terms of globalization tend to assume that this process, the inevitable expansion of markets, transformation of former comprador positions into "indigenous" capitalists, mutation of national boundary solidity, and creation process of a new supranational financial class, has fundamentally destroyed the former "Eurocentrism." But we should not be too quick to assume that the order of the world has fallen away, has given birth to something entirely new. As we can see today, the discourse of the "clash of civilizations," the substantialization of the "civilizational difference," is in fact being strongly reinforced. When we understand this "Eurocentrism" in a dynamic way, that is, not in relation to some mythic fantasy of physiological substantiality or as something located in a specific territorial configuration but rather as a *gradient of investment*, we can understand that the logic of hierarchy in the formation of bordering effects, boundaries, classifications, and so forth around specific nodules of intensity has by no means been replaced by a fluid world—rather, capital and the system of nation-states are in a process of transformation and reorganization, formulating and deploying new mechanisms of the same ideational flows that form these identificatory gradients themselves. Thus, as we have emphasized continually, this "tendential" aspect of dislocation by no means indicates that area studies has ceased to be a problem, that it is "inevitably" or "necessarily" withering away. Rather, it simply demonstrates that area studies is increasingly sundered and separated from the historical tendencies and social composition of forces that produced it in the first place. As a result, it is simultaneously *increasingly* a question of investment.

It is not that "area studies" is dead or that it must be rejected or endorsed. The very concept of "area" has ended—it has atrophied and changed its function: what can area studies do when its object, area, has detached itself from the realm of geopolitics and entered also into the realm of biopolitics? How can area studies possibly turn into a critical and transformative knowledge production when it is no longer possible to legitimate itself in terms of area? Thinking through this question from the standpoint of the constitution, maintenance, and reproduction of this intense zone of effects called "area," and attempting to negotiate its logic on the level of theory and history, can be one clue as to how we might remap ourselves and our relations, how we might sketch new "cartographies of desire" for ourselves and for the possibilities of our lives beyond the limits of area: to begin such a project is the goal of this issue of *positions: asia critique.*

Notes

We owe many thanks to Felix J. Fuchs for his editorial assistance.

1   For a discussion on the concept of the regime of separation, please refer to Sakai 2009, 2012.

2   It is well known that Benedict Anderson described the formation of the nation-state by referring to a specific form of imagined community sustained by what Walter Benjamin called "homogenous, empty time" (Benjamin [1968] 1973: 265; quoted in Anderson 1983: 24). It is noteworthy in this context that synchrony must be clearly distinguished from simultaneity since what Fabian discussed in terms of coevalness is a modality of simultaneity, while essentially synchrony is *not* about time. As Peter Osborne (1995: 27–28) has argued, synchrony is atemporal and irrelevant to time. It is not a temporal category.

3   In Johannes Fabian's analysis of the erasure of coevalness, this point is left unspecified. Anthropological discourse introduces two temporalities, but the temporality of *Western* anthropologists does not necessarily guarantee that the anthropologists and their readers of the West share a homogeneous time, even though they are manifestly *separated* from the natives of a non-Western location.

4   In contrast to the conventional comprehension of technology by which the subject manipulates and transforms the object for a predetermined objective, subjective technology implies a different conception of technology whereby the subject transforms, reconstitutes, and manufactures itself.

5   For the difference of mapping and tracing, see Deleuze and Guattari 1987: 1–25.

6   In this respect, we follow the perspective of Naoki Sakai and Jon Solomon (2006: 1–35) that directed their introduction to *Translation, Biopolitics, Colonial Difference* in which Carl Schmitt's (2006) discussion of the system of international law in *The* Nomos *of the Earth in the International Law of the* Jus Publicum Europaeum played one of the guiding roles.

7   What is suggested here is that, in the international world, its very formation through mapping is repressed and deliberately overlooked. It is as if the divisions and segmentations of the international world were simply reflective of the naturally given, as if the order of internationality were merely descriptive.

8   In their discussion of the formation of the state in primitive societies, Gilles Deleuze and Félix Guattari (1987: 432) state, "The appearance of a central power is thus a function of *a threshold or degree* beyond which what so anticipated takes on consistency or fails to, and what is conjured away ceases to be so and arrives." It is precisely in this regard that the integrity of state sovereignty and the unity of a language are comparable to one another. "Speech communities and languages, independent of writing, do not define closed groups of people who understand one another but primarily determine relations between groups who do not understand one another: if there is language, it is fundamentally between those who do not speak the same tongue. Language is made for that, for translation, not for communication" (430).

9 The conflation of Europe and the West belongs to recent history. Until the end of the nineteenth century in Western Europe the term *the West* was rarely used to connote Europe. Not only do these two geopolitical designations have many and varying denotations, they also have a wide range of variations in connotation in different languages. The terms that might be rendered as *Euro-America* and *Western Europe* are often used as synonyms for *the West* such as in the languages of Northeastern Asia. However, particularly since the Second World War, the conflation of Europe and the West has been institutionalized in discussions on cultures, civilizations, ethnicities, and races, in short, on issues pertaining to anthropological difference in general.

10 International law actually defined the procedures of land appropriation—a seizure of land whereby to give what Schmitt calls "nomos" land-based order and orientation—through which a space outside the international space was subjugated to the sovereignty of a state within the international space (see Schmitt 2003; esp. chap. 5 and part 2, "The Land-Appropriation of a New World," 80–138).

11 For an extensive development of this concept of "capture" in relation to the schema of the West and the Rest, see Walker 2018.

12 For a brief explanation about *humanitas* and *anthropos*, see Chakrabarty 1993; also see Nishitani 1998: 287–88; Sakai and Nishitani 1999: 20–22, 1038; Sakai 2000: 71–94.

References

Anderson, Benedict. 1983. *Imagined Communities: Reflections on the Origin and Spread of Nationalism*. London: Verso.

Bartholf, Mark C. 2011. "The Requirement for Sociocultural Understanding in Full Spectrum Operations." *Military Intelligence Professional Bulletin*, October–December: 4–10. www .fas.org/irp/agency/army/mipb/2011_04.pdf.

Benjamin, Walter. [1968] 1973. *Illuminations*, translated by Harry Zohn, edited by Hannah Arendt. London: Fontana.

Chakrabarty, Dipesh. 1993. "Marx after Marxism: Subaltern Histories and the Question of Difference." *Polygraph* 6, no. 7: 10–16.

Deleuze, Gilles, and Felix Guattari. 1987. "Introduction: Rhizome." In *A Thousand Plateaus: Capitalism and Schizophrenia*, translated by Brian Massumi, 1–25. Minneapolis: University of Minnesota Press.

Fabian, Johannes. 1983. *Time and the Other: How Anthropology Makes Its Object*. New York: Columbia University Press.

Foucault, Michel. 2004. *Sécurité, Territoire, Population: Cours au Collège de France, 1977–1978*. Paris: Seuil/Gallimard.

Foucault, Michel. 2007. *Security, Territory, Population*, translated by Graham Burchell. London: Palgrave.

Fukuyama, Francis. 2012. "The Future of History: Can Liberal Democracy Survive the Decline of the Middle Class?" *Foreign Affairs*, January–February. www.foreignaffairs.com/articles/2012-01-01/future-history.

Hall, Robert B., Wendell C. Bennett, Donald C. McKay, and Gerold T. Robinson. 1947. *Area Studies: With Special Reference to Their Implication for Research in the Social Sciences.* New York: Social Science Research Council.

Hall, Stuart. 1996. "The West and the Rest: Discourse and Power." In *Modernity: An Introduction to Modern Societies*, edited by Stuart Hall, David Held, Don Hubert, and Kenneth Thompson, 184–227. Malden, MA: Blackwell.

Mill, John Stuart. [1861] 1972. "Considerations of Representative Government." In *Utilitarianism Liberty Representative Government*, edited by H. B. Acton, 187–428. London: J. M. Dent.

Nishitani, Osamu. 1998. Translator's postface II to *Le crime du caporal Lortie*, by Pierre Legendre, 287–88. Kyoto: Jinmon Shoin.

Osborne, Peter. 1995. *The Politics of Time: Modernity and Avant-Garde*. London: Verso.

Patterson, William. 2016. *Democratic Counterinsurgents: How Democracies Can Prevail in Irregular Warfare*. London: Palgrave Macmillan.

Rahola, Federico. 2003. *Zone definitivamente temporanee: I luoghi dell'umanità in eccesso.* Verona: Ombre Corte.

Said, Edward. 1978. *Orientalism*. New York: Pantheon.

Sakai, Naoki. 1991. *Voices of the Past: The Status of Language in Eighteenth-Century Japanese Discourse.* Ithaca: Cornell University Press.

Sakai, Naoki. 1997. *Translation and Subjectivity: On "Japan" and Cultural Nationalism.* Minneapolis: University of Minnesota Press.

Sakai, Naoki. 2000. "The Dislocation in the West." In *Specters of the West and Politics of Translation.* edited by Naoki Sakai and Yukiko Hanawa, Vol. 1 of *TRACES: A Multilingual Series of Cultural Theory and Translation*, 71–94. Hong Kong: Hong Kong University Press.

Sakai, Naoki. 2009. "Imperial Nationalisms and the Comparative Perspective." *positions: east asia cultures critique* 17, no. 1: 159–206.

Sakai, Naoki. 2012. "*Positions* and Positionalities: After Two Decades." *positions: asia critique* 20, no. 1: 67–94.

Sakai, Naoki, and Osamu Nishitani. 1999. *Sekaishi no kaitai (Deconstitution of World History)*. Tokyo: Ibunsha.

Sakai, Naoki, and Jon Solomon. 2006. "Introduction: Addressing the Multitude of Foreigners, Echoing Foucault." In "Translation, Biopolitics, Colonial Difference," 1–35. Hong Kong: Hong Kong University Press.

Scales, Robert H. 2006. "Clausewitz and World War IV." *Armed Forces Journal*, July 1. www.armedforcesjournal.com/clausewitz-and-world-war-iv/.

Schmitt, Carl. 2006. *The* Nomos *of the Earth in the International Law of the* Jus Publicum Europæum, translated by G. L. Ulmen. New York: Telos Press Publishing.

Solomon, Jon. 2014. "Invoking the West: Giorgio Agamben's Philosophy and the Problems of Civilizational Transference." *Concentric* 40, no. 2: 125–47.

Virilio, Paul, and Sylvère Lotringer. 2002. *Crepuscular Dawn*. Los Angeles: Semiotext(e).

Walker, Gavin. 2011. "Primitive Accumulation and the Formation of Difference: On Marx and Schmitt." *Rethinking Marxism* 23, no. 3: 384–404.

Walker, Gavin. 2012a. "Citizen-Subject and the National Question: On the Logic of Capital in Balibar." *Postmodern Culture* 22, no. 3. doi:10.1353/pmc.2012.0018.

Walker, Gavin. 2012b. "Gendai shihonshugi ni okeru 'minzoku mondai' no kaiki: Posutokoroniaru kenkyū no aratana seijiteki dōkō" ("The Return of the 'National Question' in Contemporary Capitalism: New Political Directions in Postcolonial Studies"). *Shisō*, no. 1059: 122–47.

Walker, Gavin. 2016. *The Sublime Perversion of Capital: Marxist Theory and the Politics of History in Modern Japan*. Durham, NC: Duke University Press.

Walker, Gavin. 2018. "The Schema of the West and the Apparatus of Capture: Variations on Deleuze and Guattari." *Deleuze Studies* 12, no. 2: 210–35.

Racializing Area Studies, Defetishizing China

Shu-mei Shih

Introduction

The so-called crisis of area studies, perceived to be the case at the end of the Cold War for having lost its raison d'être of information retrieval for strategic purposes of the US government, and treated with "revitalizing" campaigns in the late 1990s,[1] no longer seems to be much on the minds of those who work within or without area studies. There can only be two reasons for what appears to be the passing of the crisis. Either the "revitalizing" has been so successful that it is now a transformed and different area studies, which, at its logical limit, should no longer be area studies as we know it, hence there is no more crisis. Or, the Cold War knowledge/power formation has somehow morphed into another form that continues to support and reproduce area studies to the extent that the crisis no longer exists: area studies has a new raison d'être.

positions 27:1 DOI 10.1215/10679847-7251806
Copyright 2019 by Duke University Press

On the one hand, in consideration of how most discipline-based university departments are openly critical of Cold War–style area studies scholarship and its complicity with the American government, and how discipline-based but nonetheless area-heavy scholarship in these departments has since become, albeit not completely, much more interarea or global, there seems to be some truth in arguing that area studies has changed. A better balance seems to have been achieved among the disciplines, with their universal and universalizing theories and models, and the areas, with their unique and particular contents that can only be accessed with years of language study and requisite field or archival work.[2] Scholars now study the *longue-durée* of globalization over an expansive geography, the oceanic routes of connectivity, empires that reach many lands, migration of peoples across national boundaries, economic and political behaviors of humans across times and places, but, except when these studies pertain to the West as the locus or the origin, a significant degree of area expertise is still required. It is therefore not surprising that there continue to be area specialists in most disciplinary departments, both those who have been there for a long time and new additions, even though they may no longer be called that. While there continues to be a division of intellectual labor between those who study the West and those who study the Rest, in terms of what Naoki Sakai (2010) has called the "regime of separation," the tension between disciplines and area studies has definitely abated from the time when Lucian Pye explicitly named it a "confrontation" in 1975 or when it was rephrased as that between theory (read, disciplines) and empiricism (read, area studies) and was heatedly debated in the 1990s.[3] More importantly, there appears to be a felt need for new area studies experts in the twenty-first century in a changed field of knowledge and power, which might explain why there has been a curious dearth of critical reflection on the problem of area studies for at least a decade.[4] If meeting Cold War needs mandated the study of strategically important nations and areas in the Rest, the needs of the current US empire for information and control, however repackaged, can only be more expansive, if also more diffusive, after the disappearance of the Second World and the capillary penetration of capitalism in its neoliberal stage.

On the other hand, in consideration of how professional area studies asso-

ciations continue to be well and alive, in actuality with increased participation by discipline-based scholars, institutional funding and organizational
structures remain largely unchanged (Title VI programs continue to exist),
civilizational and culturalist scholarship keeps on being produced, and
some area studies scholars are comfortably ensconced in their own separate
departments or programs, it appears that there is greater truth to the second
claim that there is now a new raison d'être for area studies, namely the new
needs of the contemporary US empire. If this is the case, we would need to
ask what has happened to the anxiety of particularity and the rhetoric of crisis? First, it seems that both the anxiety and the crisis appear to be circumscribed in scope and short-lived in duration. An argument can be made that
the Orientalist phase of area studies, identified by Sucheta Mazumdar (1991)
to be from around the late eighteenth century to the 1920s, largely following Edward Said's periodization of Orientalism, when colonial bureaucrat-
scholars and missionaries had gone to the areas out there and become clas-
sicists and philologists, has never seen a complete closure.[5] The continued,
unbroken valorization of distant texts produced by distant peoples in either
distant lands or distant times appears to have sustained the prolongation
of this phase to the present day, even as the critique of Orientalism by now
has become almost too clichéd.[6] Second, in the case of Asia, the anxiety
and crisis have been more or less thoroughly contained by the new popularity of Asian languages with the rise of Asian economies, the most recent
and overwhelmingly compelling case being Mandarin Chinese. Hundreds
of thousands of students now routinely enroll in Mandarin courses across
US institutions of higher learning and postgraduate teachers of language
and literature are needed to man these courses. Institutional funding therefore continues to flow into these programs, now justified by student interest, legitimizing traditional area studies scholarship, which provides cultural
content to otherwise "instrumentalist" linguistic acquisition for business or
other purposes of application. Furthermore, there has been more and more
Chinese funding—through Confucius institutes, Confucius classrooms, and
other forums—supporting a steady stream of current and future students
of area studies. Such Chinese funding, explicitly aimed by the Chinese government to increase "Chinese soft power while silencing or drowning out

critical views of China" (Fiskesjö 2018: 221), has naturally caused anxieties of interference, but numerous institutions of higher education continue to host these programs.

My point is that despite the flurry of scholarly activities to transform area studies—to globalize and deterritorialize it, to move it from the spatial organization of the regions or nations to oceans (see Chong 2007; Lewis and Wigen 1999), empires (see Tierney 2010; Perdue 2005), and borderlands (see Shin 2006; Giersch 2006), to name just a few—the object to be transformed, area studies, has proven to be quite recalcitrant. In fact, there seems to be new life breathed into it. The question for me is why so and how so in the post–Cold War era, and I suggest that race is the pivot around which this question can be fruitfully explored, precisely because it is one of the most, if not *the* most, "consistently repressed" problems in area studies (Sakai 2009: n.p.), which has long shown a "persistent denial of racial inequities" (Chow 2002: 108). It should not be a surprise to point out that the separation between the West and the Rest turns on racial difference, and the "Rest studies" conducted by American academics must disavow this particular difference for presumed academic respectability, intellectual neutrality, and apparent liberalism. In other words, we are, in our contemporary era of empires, living the continuous history of the world colonial turn from the late fifteenth century, when race as an organizing category was inaugurated.[7] Area studies as support for US empire, whether through functionalist social sciences or culturalist humanities, seems to continue to repress the racial logic of the empire vis-à-vis racialized populations out there in the areas as well as its internal racial minorities within the metropole. Persistently, the racial difference of others in the areas out there is framed as cultural difference, and the contiguity between those racial others and racial minorities within metropolitan centers is largely displaced, if not disavowed. As we will see below, area studies has been instrumental in the suppression of this contiguity and, in being so, has safely preserved the Rest out there and guarded it against the infiltration by the West.

Racializing Area Studies

Harry Harootunian has, perhaps more than anyone else, offered the sharpest criticism of the affective economy of area studies in terms of the area experts' relationship to the areas they study. Calling it the "Pierre Loti effect," named after the quintessential French exoticist writer, Harootunian bemoans how, more than half a century after the Second World War, American area studies experts continue to want to either "destroy" or "marry" the object of their research, with the latter being an especially pronounced desire in the post–Cold War context. Studying the areas in a particular way, Harootunian (2002: 161) argues, seems to fulfill the experts' "exotic and erotic fantasies for re-identification." Marrying native women achieves this function most expediently, as the native wives (native husbands are rare instances) can "double as native informants and stand in for the field experience in those long durations away from the field. If this often constituted a kind of passage in the rite of training, it also sealed an identity with Japan (or the area) that worked to further foreclose the possibility of critique. It became an integral part in the process of acquiring identity with the area" (162).

In the process of studying/marrying the area, scholars acquire the characteristics of the area they study, without pausing to consider "what it means to invade, inhabit, and 'snatch' the body of another" to the extent they become defenders of the area where something like patriotism or nationalism on behalf of the area occurs (166). According to Masao Miyoshi and Harootunian (2002a: 5–11) in another essay, this need for identification is the reason why area studies experts seem to take on an "ex cathedra position," judging, with near religious vigor, the loyalty and "treason" of younger scholars who enter the field, and selectively denouncing the unfaithful as "apostates" and "heretics" and threatening them with excommunication. Their defense of the area as well as arbitration of what the area must or must not mean—the two are connected and they compound each other in effect—lead to nothing less than a foreclosure of critique, that which defines what we do as critical intellectuals, unless of course criticality is not considered a requirement for academic work. In Fabio Lanza's detailed study, even for the radical group of Asianists who during the global 1960s had formed the Committee of Concerned Asian Scholars and who had been critical of the very "intellectual constitution of Asian studies as a field" (2017: 2), their "love" and "affection" (3) towards Maoist China sometimes caused them to show "glaring examples of voluntary myopia" and "huge lapses of judgment" (132). There were, in short, different reasons for loving the area, but love they/we did and do.

If area experts have exhibited an overwhelming amount of racial love toward the natives in these areas, especially toward the gentler half whom they choose to marry, one can say that there has been a proportionate expression of racial hatred toward the proponents of ethnic studies. This is especially the case for Asianists who have expressed "extreme hostility," according to Naoki Sakai, to Asian American studies and its scholars, who are mostly Americans of Asian descent. The basic irony here is that while the majority of Asianists are European Americans, their native spouses are Asians on the way to becoming Asian Americans through the inevitable processes of acculturation and citizenship in America. The conundrum here involves how one set of Asian Americans, albeit the newly minted ones, is dearly loved while another set of Asian Americans—Asian Americanists—is viewed with hostility, and how we may understand such a racial logic.

Multiple and interrelated explanations could be offered here. As I see it, one explanation concerns the difference and competition between *global* multiculturalism and *domestic* multiculturalism. By global multiculturalism, I refer to the process of how non-Western societies and nations become epitomized as particular or reified cultures, partly thanks to the work of area studies, constituting a global multiculturalism in which individual nations become individual cultures (see Shih 2007). Since nations are reduced to cultures, those who study these nations become what Rey Chow (2002: 112) has called "glorious multiculturalist[s]," who take it as their responsibility to defend the non-Western cultural traditions, especially in their unadulterated form, against the onslaught of homogenization or, worse yet, contamination brought on by Western-led globalization. This "glorious," global multiculturalist position may sometimes collapse with the Third Worldist position as they both hold romantic views of the areas they study, but it is completely opposite to the other Third Worldist position that recognizes the existence of the Third World—namely, racialized minority communities—within the First World.[8] The global multicultural area experts seek to bring scholarly knowledge about the areas and their cultures to view, where the natives are predominantly represented through cultural difference rather than racial difference.

In contrast, the domestic multicultural agenda is more than cultural recognition for its constituents, as it involves a concomitant demand for economic redistribution and political representation of minorities within the United States. Most crucially, domestic multiculturalism turns on racial difference, which, by definition and by necessity, is more confrontational as it is also more political. In this regard, Evelyn Hu-DeHart's explanation of the fundamentally divergent institutional logics of area studies and ethnic studies remains pertinent:

> Area studies programs arose out of American imperialism in the Third World and bear names such as African Studies, Asian Studies, and Latin American studies. These programs were designed to focus on US/Third World relations and to train specialists to uphold US hegemony in regions in which the US had heavy economic and political investments. Area studies scholars have become far more critical of US/Third World relations since the antiwar movement of the 1960s, and many have adopted Third World perspectives. However, they are still predominantly white male scholars entrenched in established departments, subscribing to and benefiting from traditional patterns of distributing power and rewards in the academy. (1993: 51)

Hu-DeHart continues that, on the contrary, ethnic studies has an entirely different rationale: "Ethnic studies programs, which grew out of student and community grassroots movements, challenge the prevailing academic power structure and the Eurocentric curricula of our colleges and universities. These insurgent programs had a subversive agenda from the outset; hence they were suspect and regarded as illegitimate even as they were grudgingly allowed into the academy" (51–52). In other words, area studies itself, whether global multiculturalist or Third Worldist, constitutes part of the white mainstream that ethnic studies aims to subvert. Might this be one of the reasons that area experts feel so threatened, in however artfully suppressed or displaced ways, by ethnic studies to the extent of expressing "extreme hostility"?

Obviously, there is more than the generalized institutional threat that ethnic studies or domestic multiculturalism portends for area studies and its global multiculturalist experts, otherwise it will be hard to explain the *extremity* of hostility involved. As I see it, the second explanation hinges on the unacknowledged *competition* as to who is the true multiculturalist, especially since the conflation of and confusion between area studies and ethnic studies, to the chagrin of area experts, continues to exist. Even Miyoshi and Harootunian (2002a: 4), sensing this global multiculturalist thrust in area studies, lament that "the teaching of cultures and histories may well be, as it already seems to be, assimilated into hyphenated ethnic studies programs that promise students identity and difference." The lament is that area studies is becoming like ethnic studies, or that ethnic studies will overtake area studies, even though neither of which has happened or likely will happen. Even when ethnicity and race are objects of inquiry in area studies, they are usually safely ensconced within those areas, rarely, if ever, spilling over to the United States. For area studies scholars, racial formation is seldom global.[9] Additionally, Asian history, for important reasons, continues to be taught in history or Asian studies departments, not in Asian American studies; in Asian American studies, it is the history of Asian immigration to the United States—which is American history—that is taught. Granted that the two histories intersect at select places, the emphases of the two history curricula are dramatically different. But the competition is real, even if only in perception and not in reality: who is better equipped to explain Asian culture, the Asian area experts or the Asian Americanists, who actually, or merely, *look* Asian? Which "Asia" are we talking about? The one in Asia or the one that spills over to the United States, the Asia of America, or America's Asia? So the question is also, who is the true multiculturalist, the global or the domestic multiculturalist? The global multiculturalist represents the cultures from the Rest at a safe distance whether they identify with the Rest or not, while the domestic multiculturalist represents the cultures that have moved from the Rest to the West but have since transformed to become integral to the West itself.

Further complications exist when this competition between area studies and ethnic studies is triangulated with postcolonial studies as a knowledge and institutional formation. Despite the mistaken but common conflation between postcolonial studies and ethnic studies, there is actually quite a bit of hostility expressed by the former to the latter, not to mention a serious contradiction between the two, similar to and yet different from that between area studies and ethnic studies. Anglophone postcolonial studies concerning the United States logically considers the United States as the former colony of the British Empire rather than the United States as empire, so the object of critique is Britain, not the United States, leading to the displacement of both the continuous settler colonial violence against Native Americans and the persistent racial inequality in the United States.[10] This type of postcolonial studies is clearly counter to the agenda of ethnic studies (which critiques the US empire at home), and this explains the kind of unhappy reactions from some ethnic studies scholars against postcolonial studies scholars, and vice versa. Anglophone postcolonial studies concerning South Asia similarly posits the British Empire as the object of critique and often takes the Third Worldist position that the *real* oppressed peoples are out there in South Asia, such as the tribals and the other subalterns, and are not in the inner cities or Indian reservations of the United States. Postcolonial scholars with strong Marxist persuasions, very similar to the position of area experts in this case, have thus often dismissed ethnic studies as nothing but identity politics.[11] This is despite the fact that other, older Marxists (mostly white male, in this case) have routinely thrown together postcolonial studies with ethnic studies as both playing identity politics. Terry Eagleton (2004: 21), for instance, throws both into the same dustbin called "the Cult of the Other" in his book *After Theory* and attributes the woes of contemporary depoliticization to the rise of the postcolonial studies cult. The measure of distance from ethnic studies has somehow become the litmus test to determine one's political correctness on a Marxist spectrum, even while political correctness as a concept is otherwise routinely dismissed when related to racial justice and multiculturalism.

The fact of the matter is that there is a greater affinity between postcolonial studies and area studies to the extent that we might even consider postcolonial studies as the more theoretically informed and methodologically sophisticated area studies, that is, transformed and improved area studies. They both study the areas out there—postcolonial studies examines predominantly former colonies, current colonies, and dependencies of the former British Empire, and secondarily of the French Empire—not the racialized minorities or indigenous peoples within the metropole, even though it is often the formerly colonized who become the new (im)migrants and racial minorities in the metropole and it is the indigenous peoples who continue to live in a permanent state of settler colonialism. It is suggestive, for instance, that Gayatri Chakravorty Spivak's (2003: 19) reinvigorated model of comparative literature as a discipline is to "collaborate with and transform Area Studies." The affinity clearly exists, even if traditional area experts deplore the theoretical lingo of the latter, rightly feeling threatened by a better and more sophisticated area studies model that is postcolonial studies. In this specific sense then, we may agree with Harootunian that postcolonial studies is the "true successor of area studies" (2002: 15). By now, it is common knowledge that the influence of postcolonial studies on area studies has been both intensive and extensive.

It may be stating the obvious that even postcolonial studies can be critiqued as Eurocentric in its delimitation of the objects of its critique. It privileges European colonialisms over Asian colonialisms, for instance, as if the latter do not deserve to be criticized—criticism, after all, is a form of labor that is encoded with and confers value. It may not be as obvious, in addition, that postcolonial studies also indirectly displaces the critique of the US empire in terms of its settler colonialism and internal colonialism. There are ontological reasons why indigenous studies needs to be suppressed— the indigenous people contest the sovereignty of the United States, period— hence creative rationales needed to be invented in the long and enduring postgenocide phase of settler colonialism.[12] For ethnic studies, the threat for postcolonial studies is of a qualitatively different kind and the rationales are accordingly quite different. Since ethnic studies scholars tend to be ethnic minorities themselves, they are deemed to be lacking the requisite distance from the object of study ("they study themselves"), whereby their scholarship can be dismissed as lacking objectivity as well as rigor. In this logic, distance from one's object of study is what confers objectivity and rigor. Naoki Sakai and Gavin Walker in the introduction to this issue call this logic of distance that of "remote control," which has sustained area studies for decades. This is distance, of course, that is artfully deployed, given the context of global flows of information. In this logic then, both postcolonial studies and area studies would naturally be considered superior to ethnic studies, which explains why discipline-based departments seldom hire ethnic studies scholars and why "American studies," perceived as American ethnic studies, also have a perennial problem with image, credibility, and prestige. It is useful here, therefore, to again invoke Hu-DeHart's reminder about the racialized material and institutional conditions that make the disparagement of ethnic studies inevitable, and how this relates to area studies:

> Asian American and other ethnic studies have enjoyed no history of cozy relationships with the United States government [as has area studies], and for good reason. Ethnic studies began with minority students demanding changes in the college curricula, at first simply because they wanted inclusion and representation. By this demand alone, ethnic studies students and faculty challenged the prevailing Eurocentric worldview of America, and its dominant self-image as white and democratic. From its inception, ethnic studies scholarship has been revisionist, and its goal is a re-visioning of America. Understandably, ethnic studies programs enjoy no priority with government funding agencies, have received modest support from certain private foundations, and have to fight perennially for a miniscule piece of university budgets. These facts, plus their almost exclusively minority faculties, have resulted in their low status on campuses. (1991:8)

This situation has not changed since Hu-DeHart wrote in the early 1990s. Unlike the native wives brought from the areas to the United States and safely tucked away in domestic or other nonacademic spaces, the native-looking Asian Americans have a visible presence on campuses and are challenging the authority of "whitestream" scholarship, to borrow a term from the Native American feminist Sandy Grande (2004). It is a threat posed not only by domestic multiculturalism, as discussed earlier, but also the unruly racial other who refuses to either stay away over there or stay safely in prescribed spaces while here. The ethnic studies scholar (say, Asian Americanist) who is a racialized minority, then, may be the limit of the area expert's (say, Asianist's) racial tolerance. He or she looks like a native from the area, but he or she is not a native. He or she is a racial other, and yet not totally other—he or she is also American. Not one, but two registers of the similarity/difference dialectic are activated in the Asianist's encounter with the Asian Americanist, which perhaps helps explain the knee-jerk, almost irrational dismissal of the latter by the former, even though this dismissal is couched in the rhetoric of academic objectivity and rigor.

There is still a different kind of unease already at the door, as the natives from the areas out there have themselves become area studies experts here, first in the humanities departments but increasingly also in the social sciences.[13] The opening up of China since the 1980s meant that more and more Chinese students receive postgraduate education in the United States, and some of them stay to become immigrant scholars working on China, about which they are supposed to have authentic, privileged, and firsthand knowledge. Insinuations that there are too many Chinese scholars overwhelming Chinese studies in the United States and that there needs to be a better balance (by which they mean racial balance) are part of the reaction to this new phenomenon. This is very different from the situation in Romance studies where historically the scholars have been predominantly native speakers from Europe and teaching is often conducted entirely in the Romance languages in question. Since Europe is not an area, it is free from the stigma of native representation found in area studies. Instead, native representation has been a form of prestige: native Europeans are most likely to be white and do not change the racial hierarchy of the academic institution and European languages are languages of prestige.[14] In Asian humanities, the reverse has been the case: all the teaching, except for advanced language courses, is done in English, and native representation has been proportionately smaller and mostly in the lower form of academic labor that is language instruction. The racialized division of labor consists of natives teaching labor-intensive language courses and nonnatives teaching content courses.

Again over two decades ago, Vicente Rafael (1994) wrote about how the presence of the immigrant scholar in area studies might redraw the distinctions between "us" and "them," here and there, because the natives are now among us, and how this redrawing makes "areas" contingent and unstable. Unfortunately, however, this has proven to be correct only to a very limited extent: it appears that native area experts themselves seldom, if ever, take up issues of here, even as they gradually become acculturated and acquire citizenship. The pressure to preserve the sanctity of the area out there and the authenticity conferred by their native status paradoxically work to maintain, rather than diminish, the distance between there and here. If the immigrant scholar chooses to transition, via a very difficult process of racialization, into a minority person of color in the United States, the native will no longer be just a representative of the area with a detached attitude to US racial relations but also a racialized minority, who is not transcendent from but is subjected to the usual prejudices toward nonwhite minorities. At the risk of generalizing, I think this explains why immigrant scholars teaching in English departments or American studies tend to accept their minority status in the United States and have a very different kind of relationship to the area from whence they hail, compared to immigrant scholars working in area studies who prefer the preservation of their role as native representatives from the area. What this means is that so long as the native area expert holds on to and relishes the privileged access to the area and declines to engage with the permeability between the West and the Rest, or here and there, the power/knowledge formation of area studies as we know it will remain largely unchanged. The racial logic of area studies involves multiple actors—the natives out there, the native-looking but racialized minorities here, the white area experts, the natives who become spouses, and also the natives who become area experts here—in the context of specific political, institutional, and disciplinary formations broadly outlined above. This racial logic necessarily precludes cross-racial representation: it is an open secret that there is a dearth of African American or other non-Asian minority scholars in Asian studies. Cross-minority representation simply cannot cohere with the demands of the racial logic described above.[15]

But there may still be an even newer kind of unease at the door, as changing historical circumstances are throwing up new realities our way. The rise of China compels us to recalibrate the symbolic and material meanings of here and there in an unprecedented era of contestation over old and new universalisms, which might mean, potentially, the end of China as an object of area studies.

Defetishizing China

A contrastive analogy between the immigrant author and the immigrant scholar will be instructive here as a specific illustration of how a similar logic is reproduced across the genres of creative writing and academic scholarship, especially as it pertains to China as the object of attention. On one side of the spectrum are those best-selling immigrant authors who have created a subgenre of fiction that I would call "China traumatism," peddling their insider knowledge about China, while exposing the wounds they sustained in China (spiced with requisite sex and violence). It is a classic form of fetishism with some minor variations. The writers hold onto "China" as the fetish object and lament its loss to authoritarianism—or, perhaps Oriental despotism with a variation—carrying the wounds of their loss around them, and cathecting this loss into passionate and marketable writing. "China" is necessary for these authors to sell their books, but they must vilify this China in order to sell more books: love and hatred perfectly balanced for audience or reader appeal. This phenomenon is, of course, not unique to Chinese immigrant authors, as is evidenced by the popularity of many novels belonging to the same genre about Iran, Afghanistan, or Pakistan. Simply, these novels and memoirs sell, period, as reading them gives the American readers the intellectual satisfaction, however superficial, of having acquired some knowledge about these other places and their cultures, on the one hand, and a sense of relief that their lives in the United States are superior and secure by contrast, and are at a safe distance, on the other. Reading these books provides two acts of self-confirmation in one. In the final analysis, their thankfulness toward their American life is no more sophisticated than the "starving children in Africa" syndrome. It is premised on a perfect balance between liberalism and self-confirmation, showing how liberalism of a certain kind is in the end a form of self-confirmation.

On the other side of the spectrum are those immigrant authors who consciously choose to take their postimmigration experience into account in their writings. The best example here is none other than the American Book Award–winning author Ha Jin, who decided about two decades ago that writing only about China, in the logic of the "regime of separation" mentioned earlier à la Sakai, is in the end problematic, even though his earlier works set in China are by no means sensationalist. As early as 2000, in an interview given to Asia Society, Ha Jin noted that "now the US has become the only subject that is meaningful. China has become history" (Caswell 2000). Honest to this experience, Ha Jin published his first "American novel," *A Free Life*, in 2007, which chronicles, in slowly unfolding accumulation of mundane details and Chekhov-like, small but poignant moments of revelation, the life of an immigrant family from China. No black-and-white portrayals of either China or the United States is available here, but simple narration and nuanced observations of concrete experiences that also show how China bleeds into the United States and vice versa. Ha Jin (2013) himself notes how this novel embodies his refusal to become a "cultural ambassador" like the early twentieth-century writer Lin Yutang or a "cultural peddler" like the Chinese French writer Dai Sijie. In a collection of short stories that followed, *A Good Fall* (2009), this interconnection between China and the United States is epitomized in the lives of new immigrants who are called to obligation and duty by China while struggling to find a foothold in the United States, their experience unexaggerated, unsensationalized, but also unmitigated. Blame may be distributed, no longer to the expedient and marketable China traumatism, but more evenly to a complex of factors in China and the United States, just as immigration as a lived experience is never simple or formulaic. There are, however, consequences to Ha Jin's refusal to act as a cultural ambassador or a cultural peddler. For someone who has practically swept up all the major literary awards in the United States for his novels set entirely in China, such as Hemingway and Faulkner awards not to mention the epitome of them all, the American Book Award, his award-winning streak did not extend to his books set in the United States. American readers and critics do not seem to see in what way these two books can be award worthy, because there is no ready box of Chinese traumatism (or exoticism or despotism; the list is predictably short) to put them in. By factoring in American experience, Ha Jin emphasizes the intimacy between the West and the Rest not as discrete entities but as mutually constituting and transforming realities. Ha Jin's 2014 novel, *A Map of Betrayal*, continues to address this intimacy, because without intimacy between here and there, there is no eponymous betrayal.[16] His 2017 novel, *The Boat Rocker*, is similarly a tale of betrayal. The Americanization of Ha Jin is, by choice and by assignation, the Asian Americanization of Ha Jin, which has turned out to be not a plus, but a risk, a minus.

It is these outside-the-box writers who can offer us productive, revealing, and critical analogies, one might even say allegories, of area studies, especially in terms of the fetishism of the area, and this is where the imaginative capacity of literature and its practice by writers take us a step further and deeper. In the spirit of British philosopher Iris Murdoch's (1971: 85) remark that art, especially literature and painting, "gives a clear sense to many ideas which seem more puzzling when we meet with them elsewhere, and it is a clue to what happens elsewhere," I turn to Sinophone Malaysian author Ng Kim Chew, a quintessential academic-cum-writer known for his desacralizing "bad boy" tactics (Wang 2001), who wrote what I think are several provocative allegories of area studies.[17] If, as an immigrant writer, Ha Jin's new place-based writings can be seen as an analogy to the immigrant area expert here, then some of Ng's stories offer allegories of nonnative area experts there in the field. There are two quintessential area expert figures in his stories, one a British colonial administrator-turned-writer who had previously been stationed in China and now works in colonial Malaya in a story entitled "Back Tattoo" ("Kebei"; 2001a) and a Japanese scholar who conducts fieldwork in Indonesia in a story entitled "Supplement" ("Buyi"; 2001b). In these instances, the two experts' object of fetish is the Chinese written script—the Sinitic script—which acquires near metaphysical value. To understand the fetishism of the Sinitic script not merely in terms of Orientalism and textualist classicism is very important here, because Ng's critique is intentionally multidirectional.[18]

In the context of Sinophone minority communities around the world, especially in Malaysia, it is useful to point out that as minority communities that are distinguished from the majority societies by the welding of linguistic and racial differences, the minority language in question often becomes overvalued. This is especially the case if the racialized or ethnicized group finds itself in situations of duress due to political, economic, or cultural hegemony of the majority. In the case of Malaysia where almost one third of the population is Chinese Malaysian and a high percentage of them speak different Sinitic languages (Hokkien, Cantonese, Toechew, as well as Mandarin) and where Sinophone education can be obtained all the way through college, the language(s) in question has a singular significance beyond its function as a tool of communication and expression, whether everyday or literary. We know this to be the case in Malaysia since the nation's independence from British colonialism in 1957, where the Sinophone community has been besieged with a constant sense of the threat of linguistic and cultural loss. This threat and the attendant feelings of pride and despair have engendered a predominant mood of melancholia toward the loss of their "Chinese" culture—since they live far away from China and their culture is under the threat of the dominant Malay culture—and this melancholia is then cathected to the written script that is supposed to represent their culture, even their race. Since they use a variety of spoken Sinitic languages, the standard Sinitic script as taught in Sinophone educational institutions has become the object of profound intellectual, cultural, social, and even libidinal investment to the point of the script itself becoming a fetish. The Sinitic script as a fetish thus seemingly takes on the responsibility of signifying nothing less than the cultural and racial identity of Chinese Malaysians.

In other national literary contexts, the written language acquires the fetishistic character when language becomes the object of formal experimentation as is the case in modernist and postmodernist writings in the West, but in the Sinophone Malaysian context, a very different cultural logic of the written language is at work. As literary historians and scholars have pointed out, realism has been the dominant and sanctioned mode of writing in Sinophone Malaysian literature since its inception in the early twentieth century (see, e.g., Tee 2003b: 245–51). Curiously, therefore, the function of the standard Sinitic script as the bearer of a complex of investments— nostalgia and longing for China, the resistance to siege by Malay-dominant society, and even a cultural superiority complex over the Malays—is most directly embodied in realist fiction wherein literature's function is supposed to reflect the authentic reality of Chinese Malaysians. In other words, language becomes a fetish not necessarily in the (post)modernist form as is the case elsewhere, but rather in the realist form, and this divergence foregrounds for us how the threat of loss produces language fetishism regardless of literary mode or form.

In Sinophone Malaysian literature, another particular tendency noted by scholars is the desire to acquire a form of classicism in the Chinese mode: the appropriation of literary and lyrical language of classical Chinese literature as an expression of the longing for the root of their being—"China"—can be found in writers across the ideological spectrum of both the Left and the Right. Efforts are made to acquire classical prose style and diction, learning it by the books, literally, as it is not the language of the everyday. The works of such writers as Tan Boon Hong (known by his pen name Sha Qin) and Lim Choon Bee have been analyzed in terms of their classical prose, hence Tan Boon Hong's statement that the Sinitic script is "the most important key to opening history" (Huang 1999: 225–26). The case of Li Yung-p'ing's quest for "pure Chinese writing," though in this case not classical lyricism but what he considers to be pure vernacularism, and the case of the Sirius Poetry Society (tianlangxing shishe) founded by Woon Swee Tin that expressed a long-distance Chinese cultural nationalism are also often used to illustrate this longing.[19] In all of these cases, the fetishization of the Sinitic script results not from the desire for formal experimentation but a profound melancholy at the loss of "Chinese" culture and language. This particular fetish of language in the realist tradition therefore has its genesis in the social, not the textual; in this sense, it is paradoxically truer, in almost a literal sense, to the demands of realist verisimilitude. Because presumed authentic classicism or pure vernacularism needed to be *learned* by effort, furthermore, due to what Kim Tong Tee (Tee 2003a) has rightly called the "deterritorialization" of the Sinitic script outside China proper, realist writing itself must by definition be full of artifice. Whether the search is for classicism or vernacularism, the script acquires a fetishistic character. The fetish as a substitute for the lost object parallels realism as a substitute for another lost object—modernism—hence the perennial debate in Sinophone Malaysian literature about the viability of modernism as a mode of writing. As is well known, the word *fetish* itself is derived from the Portuguese term *feitiço*, meaning "artificial, skillfully contrived," and the word in turn comes from the Latin *facticius*, meaning "made by art" (Bernheimer 1993: 63). By inference, when the Sinitic script is turned into a fetish, the resultant writing will naturally be even more artificial and skillful, that is, artful and artistic, and this may be why there is such a rich body of astonishingly beautiful and hauntingly evocative literary works penned by Sinophone Malaysian writers.

Art historian Hal Foster (1993) reminds us that the term *fetish* has three uses, the anthropological, the Marxian, and the Freudian. To give a short summary here: In the anthropological usage, the fetish is a thing possessed of a (super)natural quality or force, as in a charm. In the Marxian sense, it refers to the reification of commodities wherein commodities become merely things alienated from human relations and human production, their exchange value having overtaken their use value, which then acquires for them a kind of mythical character. In the Freudian sense, the fetish is the substitute object on which the emotional and libidinal energy of the fetishist is concentrated so that his castration anxiety can be displaced and his ego restored. Note that the Freudian fetishist is almost always male and fetishism has a great deal to do with his sense of manhood or masculinity. In all of these three uses, the fetish is the object of "overvaluation," concerning the realms of the spiritual, the economic, and the erotic.

In the stories of Ng Kim Chew, the fetishism of the Sinitic script is almost a textbook case for illustrating the many dimensions and functions of the fetish. One Chinese Malaysian fetishist obsessively writes the Sinitic script then burns the writing, mixing "fire, blood, sweat, and semen" in the process (Ng 1997b: 88). Another fetishist's life's purpose is to be able to carve his name in the Sinitic script on some rock in "fatherland" China, his long-distance nationalism leading him to sacrifice his life to the borrowed cause of the Chinese Communist Party against the Japanese invaders who were occupying Malaya during World War II (Ng 1997a). The most scandalous cases, however, are reserved for non-Chinese Malaysians. The protagonist of "Back Tattoo," a story set in colonial Malaya, is a British colonial administrator-writer who had once served in China, married a Chinese woman, and now works in Malaya. Like many of Ng's fetishists, he has discovered the absolute beauty of the Sinitic script, succumbed to it as if it were a "charm," and vowed to write his modernist masterpiece as powerful as James Joyce's *Ulysses* in this script. He proceeds to wed the medium of writing—the Sinitic script—to a nonprint canvas that would best capture the essence of modernist writing as a critical meditation on the (im)permanence, ephemerality, and fragmentation of representation. His Chinese wife, whom he brought over to Malaya, recalls: "He has finally found the irreplaceable, revolutionary modernist form for writing in the Sinitic script: using the most modern script, a live medium, immediate publication, shortness of fleeting life—this fleeting moment gestures toward *dasein*, the absolutely untranslatable one-time-ness, with absolutely no copies, and thoroughly transcending the lifeless medium of writing the Sinitic script on paper or paper-like objects. . . . He felt that he has captured the most secret meaning of the script [in this way]. (2001a: 353).[20]

This medium he discovered is the human skin, the backs of Chinese coolies brought over by the British colonizers, onto which he proceeds to inscribe his masterpiece in the Sinitic script. Using all kinds of unseemly methods, including purchasing coolies and locking the disobedient ones in cages, he tattoos onto their backs in blue ink and sometimes even rapes them. Here, modernist formal experimentation and fetishism of language are taken to their Orientalist extremes, revealing the dark underside of what passes in the name of artistic creation in this case: sadism and rape. The coolies whose backs are inscribed are thus symbolically castrated, forever living in deep shame, not to mention the shame of actual rape for some of them; in the process, the Orientalist modernist's potent sexuality (he also ravages through the bodies of local prostitutes and his wife) is revealed. Here, the Sinitic script takes on anthropological, mythic, and erotic dimensions of the fetish all at once, and the fetishist's masculinity and ego are more than amply confirmed.

More surprise awaits, however, when we learn toward the end of the narrative that the Sinitic characters carved onto the backs of the coolies are mostly malformed, sometimes with one character splitting into several characters, and sometimes with several characters combining into one character. There is a sly reference here to Ezra Pound's dismantling of the Sinitic script in a similar way in his creative translation of Chinese poetry, following the call by Ernest Fenollosa in the latter's famous article "The Chinese Written Character as a Medium for Poetry" (first draft 1906; see Fenollosa and Pound 2008), for which Pound served as editor. As is the case with Orientalism, the fetishized object does not need to be correct or authentic, as long as it carries the fantasies and investments of the fetishist adequately. Better still, the fetishized object is necessarily the mistake. The British protagonist in "Back Tattoo" has been looking for China in all the wrong places: despite marrying a Chinese wife, he is after all living in Malaya; despite his devotion to the Sinitic script, his act of inscription turns out to be a form of violation and the writing itself turns out to be nothing but nonsensical gobbledygook.

The other prominent fetishist can be found in Ng's (2001b) story "Supplement." Takatsu is a Japanese scholar whose obsession is to collect any "remains" of the famed Chinese writer Yu Dafu, who was exiled to Malaya (1938–42) and Indonesia (from 1942 until his death) and presumed to have been executed by the Japanese military in Sumatra at the end of the Pacific War in 1945. While the anti-Suharto, anti-Chinese riot was raging through the streets of Jakarta in May 1998, Takatsu leads a crew of Taiwanese documentary filmmakers, who had just completed a documentary on Yu Dafu, on a journey to prove his hypothesis that Yu had never been killed but had lived anonymously for many years beyond the war. The context for such a journey is an actual, considerable Yu Dafu mania among scholars and readers in Japan, and a deep, complex fascination for him in Sinophone literary circles in postcolonial Malaysia and Singapore.[21]

Ng has written, for instance, no less than three stories working out different scenarios of Yu's life beyond his presumed execution.[22] In the story, "Supplement," we have the most Orientalized figure of Yu, appearing as a ghostly, finely wrinkled centenarian in a flowing antiquarian Chinese robe and surrounded by dozens of descendants all dressed in traditional Chinese costumes living on an isolated island guarded by pirates. Takatsu and the Taiwanese filmmakers meet this Yu upon being stopped and abducted by the guard-pirates during their boat journey following some credible clues in search of Yu. They are escorted to the large central hall of a traditional, four-wing compound that looks like "the stage set for a traditional costume movie" and is filled with dense aromas of cinnamon like "an old Chinese pharmacy." All of their belongings are searched, revealing the content in Takatsu's two suitcases: a large quantity of the so-called Banana banknotes—because there is a picture of banana trees on one of the denominations—printed by the Japanese military in Southeast Asia during World War II. On these Banana banknotes were inscribed, in fragments, four novels supposedly handwritten by Yu, hence Takatsu's frantic cry as they were put to the pyre. Takatsu's fetish objects—Yu's writings—are lost, his clothes are slashed, and his body is strip-searched, including his anus.

The next day Takatsu and the Taiwanese filmmakers were let go, and they return to their respective countries. A few months later, the filmmakers receive a letter from Takatsu, who claims that he has successfully transported Yu's corpse, sans his penis and testicles, to Japan, after his tomb was discovered and excavated. He details in the letter that the well-preserved corpse exudes strong aromas of cinnamon, star anise, and wild pepper.[23] He says that he is still looking, most earnestly, to retrieve Yu's penis and testicles—the "three treasures"—no doubt taken by some Chinese pharmacists and being soaked in wine for their medicinal benefits. Rumor has it also that an old, venerable Taiwanese politician's two amber-colored balls used for hand exercises look very much like his testicles, and an old woman's legendary pipe perched constantly in her mouth looks very much like his penis. This is where the story ends. Now the fetishist of writing has officially transitioned to the fetishist of body parts, especially sexual organs.

Generally speaking, we can say that there are two different kinds of fetishists in Ng's stories: Chinese Malaysians who longingly fetishize the Sinitic script as a synecdoche to the far-away, imagined motherland of China, and the Orientalists looking for the same fetish and its macabre substitutes. If the former may be allegories of the immigrant area expert or self-ethnographer, the latter may be extreme allegories of overzealous area experts. For both, their intellectual and libidinal investments are interlinked and their fetishism is attached to a displaced, incomplete, or even wrong object. After all, they all one way or another cathect their longing to the mistake, as they were trying to find China in all the wrong places. The main disjunction lies in the fact that the Sinitic script is taken as a synecdoche where the part is geoculturally distant from the whole (China), and they will never be able to bridge the impossible distance between Malaysia and China, hence melancholia, hence fetishism. Distance, it turns out, is a fetishism-producing mechanism or a structure that is conducive to producing fetishism.

But the fact of the matter is that, as numerous Sinophone Malaysian writers have shown us, their "China" is right in their midst: for the Chinese Malaysians, their reality is the one on the ground, where their Chineseness is mixed up with, in myriad ways, everything else that is on the ground, not the reified, pure, authentic Sinitic script that is not violated or influenced by the livedness of their Malaysian experience. Numerous stories by other Sinophone Malaysian writers, most notably Li Zishu and Chang Kuei-hsing, would express, in deeply allegorical ways, the imperative of such a place-based perspective.[24] Similarly, for the British colonial administrator in "Back Tattoo," a place-based perspective would mean an entirely different kind of attention to the Chinese minorities—the coolies—in Malaysia. Indeed, it is Chang's novels set in the Borneo rainforest that would offer us a dark and violent counter-epic of Chinese settlement in Malaysian Borneo beginning with the generation of Chinese coolies in the late nineteenth century. His version is not the instrumentalization and dehumanization of coolies as merely surfaces to write on, but coolies as historical subjects who cleverly manipulated the relationship between the British colonizers and the indigenous Dayaks to become plantation owners who exploited land and nature, as well as the indigenous people.[25]

If Ng's stories seem all too fantastical to be convincing, there had been numerous records of exploits by historical personalities such as prominent writers from the colonial period who tried to capitalize on the distance between the here and the there. The case of André Malraux (1901–76) is probably one of the most scandalous, hence worth a brief mention. When the French writer was young, he stole stone bas-reliefs weighing eight hundred kilograms from Banteay Srei, which is northeast of Angkor Wat in Cambodia, to sell them for profit back in France. He was arrested and only narrowly avoided a long prison sentence, thanks to the concerted efforts of prominent French writers who petitioned on his behalf. He would go on to write a novel entitled *The Way of the Kings* (*La voie royale*; [1930] 2005), where he turned this episode of his life searching for and stealing the carvings into an existential quest for the meaning of life, expressing a strange sincerity of purpose that is utterly devoid of any hint of remorse or guilt. Like the coolies who are treated like beasts in Ng's story, Malraux depicts, without irony, the natives of Cambodia as subhuman and living a "malarial life" in a fetid place that is decaying and rotting (Malraux 2005: 41). For latter day readers, there is nothing surprising about the turning-animal of natives and the primitivizing of the distant land, which are typical tropes of colonial narratives in the Orientalist vein. But the surprise is that, the stealing of eight hundred kilograms of carvings, which was a crime under French law, also did not seem to have compromised his career but rather to have set the stage for him to become an illustrious cultural figure in France, including a long stint as the minister of culture.

Three years after the publication of *The Way of the Kings*, Malraux would publish his most celebrated novel, *Man's Fate* (*La condition humaine* 1933), which is set in China. The novel garnered quite a bit of credibility for its depiction of the 1927 uprising against the Kuomintang by an international cast of leftists in Shanghai, thanks to the dubious claim that the author participated in the revolutionary movements in China between 1923 and 1936.[26] His claim to firsthand, participatory knowledge about China turned out to be more than helpful to his career. He received the Prix Goncourt for the novel, even though this claim later turned out to be false. Luckily for Malraux, because the novel is a work of fiction, it remained a classic, not as a credible account of the transpired events, but as a text that has something to offer in terms of reflections on death, life, and love during a critical period of decisive human action such as a political insurgency. By then, Malraux had already capitalized on the distance between China and France and his "expert" knowledge of China, which had given him not only the literary but also the financial breakthrough that he had sorely needed. He stole the eight-hundred-kilo stone reliefs for lack of money, after all: the audacity of stealing such astonishingly heavy objects was probably equal to the extreme severity of his destitution at the time.

Ultimately, the simple fact is that the China imagined from Malaysia, France, and the United States, is an imaginary China. For Ng, the further complication is that he actually wrote these stories as a long-term resident alien in Taiwan.[27] Ng's own relationship to his homeland, Malaysia, itself replicates, with some difference, the similar dynamic of a Chinese Malaysian's relationship to China (see Chong 2006). The multiple mediations of geographical dislocation that we must factor into our understanding of any work of representation, including academic scholarship, must be calibrated with place-based perspectives to arrive at a necessary dialectic between there(s) and here(s). This is especially the case for Chinese studies scholars now with the rise of China in the global scene, vying for not only political and economic power but also the power of representation. Before we know it, the Gramscian war of position is already upon us as to who has the right to represent China and who has the power of adjudication as to what and what not to say and write about China. The Chinese state-sponsored critique of New Qing History produced in the United States is a case in point.[28] This is especially telling, in relation to the question of ethnicity (and race by implication), since New Qing History is almost the only subfield in Chinese studies, besides anthropology, that has foregrounded issues of ethnicity. Could it be that racializing Chinese studies—that is, to defetishize China—is also threatening to the Chinese state?

Racializing Area Studies, Again

The presumed oxymoronic relationship between race and area studies, as I have tried to show, is born out of multiple, overlapping genealogies that have reinforced each other: the textualist, Orientalist tradition; the Cold-War mandate; the racial politics of institutional organizations; and the stubborn scholarly practice of the disavowal of race and the racial reality within the United States. And now we must also add the desires of those regimes in the areas themselves that prefer a deracialized area studies so that their own racial hegemony—such as Han hegemony over the other races and ethnicities in China—can be maintained. After all, W. E. B. Du Bois's ([1903] 1995: 4) trenchant statement that "the problem of the Twentieth Century is the problem of the color line" can be ignored only if one thinks one has a reason or justification for doing so. Up to now, area studies offered a convenient excuse.

There are two related problems that need to be overcome before a different area studies that is not afraid of its racial unconscious can emerge. The obvious problem is the need to study race and ethnicity critically in area studies contexts. The second problem is how when race is actually foregrounded as an issue over there, we need to bring it over here and set the two in active confrontation and dialogue, that is, in a relation. The sum of relations unsettles the unacknowledged hierarchy of subjects and objects of knowledge and the artificial division between there and here. Most importantly, these relations call scholars into ethical reflection on their relationship to racialized minorities in their midst. After all, the US-born and immigrant minority communities are themselves produced by political, historical, and economic processes interlocking the United States and those areas from which the minorities and their ancestors hail, just as the particular, historically determined European immigration to the United States helps explain why the study of Europe can be universalized and has always escaped the logic of area studies. Racialization is relational and global at the same time, and it is always already comparative as a historical as well as psychosocial process.[29] The comparative approach to a comparative process therefore seems to be only logical, if belated. A critically considered comparative approach, I would also like to argue, is an ethical demand that helps stave off the reification and consumption of the other and the attendant universalization and artificially produced neutrality of the self. Both are complex and fragile beings implicated within related formations of power and knowledge, which are also racial formations. The tension between area studies and ethnic studies becomes moot once their interlocking racial formation is no longer disavowed. This is what I mean by the racialization of area studies.

Notes

Earlier versions of this article were delivered as lectures at the University of Washington, Duke University, and the National University of Singapore. I am grateful to the audiences at these occasions who responded to the lectures, for the comments by the two anonymous readers for *positions*, and to the coeditors of this issue.

1 The reference is to the two revitalizing area studies campaigns launched by the Social Science Research Council and the Ford Foundation in 1996 and 1997, respectively. The latter campaign carried the word *revitalizing* in its title.

2 Pheng Cheah (2001) similarly describes the case of disciplines versus area studies as constructed to be that between the universal and the particular.

3 Lucian Pye (1975) named this conflict in an eponymous essay entitled "The Confrontation between Disciplines and Area Studies." In the 1990s, the disciplines/theory versus area studies/empiricism debate was revived again on the pages of *Modern China* for the field of Chinese history (see Huang 1991, 1993), coinciding with a similar debate among scholars in Chinese literary studies articulated as "Western theory versus Chinese reality" (see Zhang 1992, 1993).

4 It is not that the usual problems of area studies have all been resolved, but that the problems are no longer brought up for critical reflection, hence the particular significance of this special issue of *positions*.

5 Edward Said himself equates the earlier mode of Oriental studies with the Cold War mode

of area studies in his discussion of the "Orientalist in his new or old guise" in "Oriental studies or area studies" in academia (1978: 2).

6 For Rey Chow's earlier critique of the Orientalism in Chinese studies scholarship, see Chow 1993.

7 For discussion of the colonial turn and racialization, see Shih 2008.

8 See, for instance, the work of Chandra Mohanty, especially the articles by Mohanty and others collected in *Third World Women and the Politics of Feminism* (Mohanty et al. 1991).

9 The most notable exception is Takeshi Fujitani's *Race for Empire* (2011).

10 *The Empire Writes Back* (Ashcroft, Griffiths, and Tiffin 1989), which has gone into sixteen reprints and is considered by many a classic in the field, is symptomatic of a parallel phenomenon pertaining especially to Australia, where the privileged colonial relationship is that between the British (as the colonizer) and the white settlers (as colonized), not that between the white settlers (as colonizers) and the aboriginals (the colonized). The confrontation and negotiation between British colonialism and settler colonialism in theoretical and historical terms was never fully accounted for in the book.

11 As Gayatri Chakravorty Spivak (2003: 28) says of ethnic studies, very problematically, in a different context, "The confrontation of old Comparative Literature and Cultural/Ethnic Studies can be polarized into humanism versus identity politics."

12 Lorenzo Veracini (2010) calls these rationales techniques of "transfer" in his very useful book *Settler Colonialism: A Theoretical Overview.*

13 The sheer quantity of primary texts one has to master in the humanities naturally advantages the native speaker turned native-informant-cum-area-expert. In the social sciences, due to the emphasis on theoretical models (all in English), native speakers who are immigrants tend to be disadvantaged. But the natives are learning theories really well too and are competing for jobs in discipline-based social sciences. Could it be that this is an aspect of the racial unconscious of social sciences when it comes to its emphasis on disciplinarity over area expertise, to prevent the native-turned-area experts from overwhelming their departments, as it does native-looking ethnic studies scholars?

14 Spanish is dealt a different fate these days due to its association with Mexico and Mexican immigrants instead of the "desired" peninsular association, but French continues to be the language of high culture and the native French scholars embodiments of that high culture. The image persists even as European language departments have dwindling student enrollments and fewer and fewer faculty positions.

15 The only exception is the study of China-Africa relations, which has in recent years become a hot topic, where one or two African American scholars are now able to or allowed to partake of Chinese studies as such. The irony here is that African Americans are not Africans, which again shows the simplistic assumption of correspondence between geography and race.

16 It is revealing that most of Ha Jin's works set in China are placed under the general category

called "Books" on Amazon's "Best Seller Rank," whereas *A Map of Betrayal*, which is about a Chinese spy in the United States who has divided loyalties to both China and the United States and told from the perspective of the second-generation Chinese American daughter, is specified as "Asian American" under the "Literature and Fiction" category. Being presumed a Chinese national writing about China makes Ha Jin a writer who can represent China, be its representative, and offer an authentic and reliable account of the distant land of China, thus acceptable and unthreatening. But the moment he writes about the United States, especially about the highly sensitive issue of spying, he becomes a minority subject to be named thus and accordingly categorized. It does not take much to see a similar logic here replicating that between area studies (Ha Jin as a Chinese writer) and ethnic studies (Ha Jin as a Chinese American writer).

17   Ng's bad boy tactics in writing usually take on a deconstructive mode and are highly productive, but his bad boy tactics in real life usually misfire and are at odds with his creative work, such as the stories discussed here. Living in Taiwan, he refuses commitment either to Taiwan (since it is Taiwan-centric) or to Malaysia (since it is Malay-dominant) and takes an exclusivist diasporic-exilic position, full of masculine resentment that explodes at whatever offends his peculiar sensibility and fragile masculinity. Such a contradictory relationship between the writing and the writing self is not uncommon, of course.

18   On multidirectional critique of Sinophone writing, see the conclusion to Shih 2007: 183–92.

19   See Ng's (1998) book of critical essays, *Sinophone Malaysian Literature and Chineseness*, esp. chaps. 3, 4, and 5.

20   The story is collected in *Dari Pulau Ke Pulau* (*You dao zhi dao*). The title of the collection is in Malay and it means "From Island to Island." All translations from the original are mine.

21   There is currently also a well-heeled Yu Dafu Society in the United States.

22   See Groppe 2010 for a detailed analysis of all three stories.

23   At this point, the author leaves it ambiguous whether the old man they met on the island is the same as the dug-up corpse, except the hint of the dense aroma of cinnamon.

24   For Li Zishu, see her award-winning, allegorical short story (2010), "The Frontier in the Nation's North" ("guobei bianchui"), in which all male members of a Chinese Malaysian family suffer from a curse and cannot live beyond age thirty unless they ingest the root of a mystical plant. The root of the plant, called dragon tongue (*longshe*), is supposed to embody the essence of China and will allow the men to survive the curse, but the plant turns out to be rootless. To consider dragon tongue/China as root, the story implies, is completely futile, since the plant is rootless. If you are forever searching for it, you will die an untimely death or live a death-in-life existence. It is a form of refusal of life on the ground.

25   For an analysis of Chang's novelistic representation of three generations of Chinese settlers in the Borneo rainforest, see Shih, 2013a. Chang continues to add to this Borneo series with his latest novel, *Wild Pigs Crossing the River* (yezhu duhe, 2018), which focuses more on the Borneo under Japanese occupation during World War II.

26 As claimed in the biography page of the novel. The novel was translated into English in
     1934, and the version I am using here is the 1990 reprint of the English translation by
     Vintage.

27 For the complex dynamics involved in the double diasporic status of Sinophone Malaysian
     writers living in Taiwan, see Chiu 2008. In my analysis, the only constant in this doubly dia-
     sporic poistion for someone like Ng is the imaginary "China" as the object of fetish, which
     explains the strange politics of Ng Kim Chew, who is completely unsympathetic to local
     causes in Taiwan and takes a strangely "superior," China-centric position when it comes to
     Taiwan. Underlying his deconstruction of China-fetishism is a paradoxical pleasure derived
     from the fetishism itself. This pleasure is irrational, hence the explosive character of his
     writing and his personal attacks on others who might offend him in any way.

28 Chinese censorship of Chinese studies scholarship around the world has been increasing in
     intensity and scope in recent years, including the various controversies involving Cambridge
     University Press, Springler, and Confucius Institutes and the denial of visas to enter China
     and Hong Kong, not to mention silent self-censorship by scholars themselves.

29 I develop this argument in Shih 2008. An abridged version of that essay appeared as "Com-
     parative Racialization," in *A Dictionary of Cultural and Critical Theory* (2010).

References

Ashcroft, Bill, Gareth Griffiths, and Helen Tiffin. 1989. *The Empire Writes Back: Theory
and Practice in Post-colonial Literatures*. New York: Routledge.

Bernheimer, Charles. 1993. "Fetishism and Decadence: Salomé's Severed Heads." In *Fetishism as Cultural Discourse*, edited by Emily Apter and William Pietz, 62–83. Ithaca, NY:
Cornell University Press.

Caswell, Michelle. 2000. "An Interview with Ha Jin." Asia Society (educational organization). www.asiasociety.org/interview-ha-jin (accessed February 12, 2017).

Chang, Kuei-hsing. 2018. *Wild Pigs Crossing the River* (*Yezhu duhe*). Taipei: Linking Books.

Cheah, Pheng. 2001. "Universal Areas: Area Studies in a World in Motion." In *Specters
of the West and Politics of Translation*, edited by Naoki Sakai and Yukiko Hanawa. Vol. 1
of *TRACES: A Multilingual Series of Cultural Theory and Translation*, 37–70. Hong Kong:
Hong Kong University Press.

Chiu, Kuei-fen. 2008. "Empire of the Chinese Sign: The Question of Chinese Diasporic
Imagination in Transnational Literary Production." *The Journal of Asian Studies* 67, no. 2:
293–620.

Chong, Fah-hing. 2006. *Guojia wenxue: Zaizhi yu huiying* (*National Literature: Hegemony
and Response*). Selangor, Malaysia: Mentor.

Chong, Terence. 2007. "Practicing Global Ethnography in Southeast Asia." *Asian Studies Review* 31: 211–25.

Chow, Rey. 1993. *Writing Diaspora: Tactics of Intervention in Contemporary Cultural Studies*. Bloomington: Indiana University Press.

Chow, Rey. 2002. "Theory, Area Studies, Cultural Studies: Issues of Pedagogy in Multiculturalism." In *Learning Places: The Afterlives of Area Studies*, edited by Masao Miyoshi and H. D. Harootunian, 103–18. Durham, NC: Duke University Press.

Du Bois, W. E. B. [1903] 1995. *The Souls of Black Folk*. New York: Signet.

Eagleton, Terry. 2003. *After Theory*. New York: Basic.

Fenollosa, Ernest, and Ezra Pound. [1906] 2008. *The Chinese Written Character as a Medium for Poetry*. Edited by Haun Saussy, Jonathan Stalling, and Lucas Klein. New York: Fordham University Press.

Fiskesjö, Magnus. 2018. "Who's Afraid of Confucius? Fear, Encompassment, and the Global Debates over Confucius Institutes." In *Yellow Perils: China Narratives in the Contemporary World*, edited by Frank Billé and Sören Urbansky, 221–45. Honolulu: University of Hawai'i Press.

Foster, Hal. 1993. "The Art of Fetishism." In *Fetishism as Cultural Discourse*, edited by Emily S. Apter and William Pietz, 251–65. Ithaca, NY: Cornell University Press.

Fujitani, Takeshi. 2011. *Race for Empire: Koreans as Japanese and Japanese as Americans during World War II*. Berkeley: University of California Press.

Giersch, C. Patterson. 2006. *Asian Borderlands: The Transformation of Qing China's Yunnan Frontier*. Cambridge: Harvard University Press.

Grande, Sandy. 2004. *Red Pedagogy: Native American Social and Political Thought*. Lanham, MD: Roman and Littlefield.

Groppe, Alison M. 2010. "The Dis/Reappearances of Yu Dafu in Ng Kim Chew's Fiction." *Modern Chinese Literature and Culture* 22, no. 2: 161–95.

Ha Jin. 2007. *A Free Life*. New York: Vintage.

Ha Jin. 2009. *A Good Fall*. New York: Pantheon Books.

Ha Jin. 2013. "Exiled to English." *Sinophone Studies: A Critical Reader*, edited by Shu-mei Shih, Tsien-hsin Tsai, and Brian Bernards, 117–24. New York: Columbia University Press.

Ha Jin. 2014. *Map of Betrayal*. New York: Pantheon.

Ha Jin. 2016. *The Boat Rocker*. New York: Pantheon.

Harootunian, Harry. 2002. "Postcoloniality's Unconscious/Area Studies' Desire." In Miyoshi and Harootunian, eds., *Learning Places*, 150–74. Durham, NC: Duke University Press.

Huang, Philip. 1991. "The Paradigmatic Crisis in Chinese Studies." *Modern China* 17, no. 3: 299–341.

Huang, Philip, ed. 1993. "'Public Sphere'/'Civil Society' in China? Paradigmatic Issues in Chinese Studies, III." Special issue, *Modern China* 19, no. 2.

Huang, Wanhua. 1999. "Discursive Practice of the Young Generation of Mahua Writers" ("Mahua xin shidai de huayu shijian"). In *World Huawen Literature in Cultural Transition* (*Wenhua zhuanhuan zhong de shijie huawen wenxue*), 224–33. Beijing: Zhongguo shehui kexue chubanshe.

Hu-DeHart, Evelyn. 1991. "From Area Studies to Ethnic Studies: The Study of the Chinese Diaspora in Latin America." In *Asian Americans: Comparative and Global Perspectives*, edited by Shirley Hune et al., 5–16. Pullman, WA: Washington State University Press.

Hu-DeHart, Evelyn. 1993. "The History, Development, and Future of Ethnic Studies." *Phi Delta Kappan* 75, no. 1: 50–54.

Lanza, Fabio. 2017. *The End of Concern: Maoist China, Activism, Asian Studies*. Durham, NC: Duke University Press.

Lewis, Martin W., and Karen Wigen. 1999. "A Maritime Response to the Crisis in Area Studies." *Geographical Review* 89, no. 2: 161–68.

Li, Zishu. 2010. "The Frontier in the Nation's North" ("Guobei bianchui"). In *Wild Bodhisattva* (*Ye pusa*), 17–40. Taipei: Ryefield.

Malraux, André. [1934] 1990. *Man's Fate*, translated by Haakon M. Chevalier. New York: Vintage. Originally published as *La condition humaine* (Paris: Gallimard, 1933).

Malraux, André. [1930] 2005. *The Way of the Kings*. Translated by Howard Curtis. London: Hesperus. Originally published as *La voie royale* (Paris: B. Grasset).

Mazumdar, Sucheta. 1991. "Asian American Studies and Asian Studies: Rethinking Roots," in *Asian Americans: Comparative and Global Perspectives*, edited by Shirley Hune, 29–44. Pullman, WA: Washington State University Press.

Miyoshi, Masao, and Harry Harootunian. 2002a. "Introduction: The 'Afterlife' of Area Studies." In Miyoshi and Harootunian, eds., *Learning Places*, 1–18. Durham, NC: Duke University Press.

Miyoshi, Masao, and Harry Harootunian, eds. 2002b. *Learning Places: The Afterlives of Area Studies*. Durham, NC: Duke University Press.

Mohanty, Chandra Talpade, Ann Russo, and Lourdes Torres, eds. 1991. *Third World Women and the Politics of Feminism*. Bloomington: Indiana University Press.

Murdoch, Iris. 1971. *The Sovereignty of Good*. New York: Schocken.

Ng Kim Chew. 1997a. "Fish Remains" ("Yuhai"). In *Dark Night (Wuanming)*, 251–78. Taipei: Jiuge chubanshe.

Ng Kim Chew. 1997b. "In the Depth of the Rubber Forest" ("Jiaolin shenchu"). In *Dark Night (Wuanming)*. Taipei: Jiuge.

Ng Kim Chew. 1998. *Sinophone Malaysian Literature and Chineseness (Mahua wenxue yu zhongguo xing)*. Taipei: Yuanzun wenhua chubanshe.

Ng Kim Chew. 2001a. "Back Tattoo" ("Kebei"). In *Dari Pulau Ke Pulau (You dao zhi dao)*, 325–59. Taipei: Rye Field.

Ng Kim Chew. 2001b. "Supplement" ("Buyi"). In *Dari Pulau Ke Pulau*, 267–90.

Perdue, Peter. 2005. *China Marches West: The Qing Conquest of Central Eurasia*. Cambridge, MA: Harvard University Press.

Pye, Lucian. 1975. "The Confrontation between Disciplines and Area Studies." In *Political Science and Area Studies: Rival or Partners?*, edited by Lucian Pye, 3–22. Bloomington: Indiana University Press.

Rafael, Vicente. 1994. "The Cultures of Area Studies in the United States." *Social Text* 41: 91–111.

Said, Edward. 1978. *Orientalism*. New York: Pantheon Books.

Sakai, Naoki. 2009. "Asia and Image of the World: Co-Figuration of the West and the Rest." Paper presented at the Global Asia in Historical Perspectives conference, Pennsylvania State University, State College, PA, October 23–24.

Sakai, Naoki. 2010. "From Area Studies Toward Transnational Studies." *Inter-Asia Cultural Studies* 11, no. 2: 265–74.

Shih, Shu-mei. 2007. *Visuality and Identity: Sinophone Articulations across the Pacific*. Berkeley: University of California Press.

Shih, Shu-mei. 2008. "Comparative Racialization: An Introduction." In "Comparative Racialization," edited by Shu-mei Shih. Special issue, *PMLA* 123, no. 5: 1347–62.

Shih, Shu-mei. 2010. "Comparative Racialization." In *A Dictionary of Cultural and Critical Theory*, edited by Michael Payne and Jessica Rae Barbera, 143–49. 2nd ed. London: Wiley-Blackwell.

Shih, Shu-mei. 2013. "Comparison as Relation." In *Comparison: Theories, Approaches, Uses*, edited by Rita Felski and Susan Friedman, 79–98. Baltimore: Johns Hopkins University Press.

Shin, Leo K. 2006. *The Making of the Chinese State: Ethnicity and Expansion on the Ming Borderlands*. Cambridge: Cambridge University Press.

Spivak, Gayatri Chakravorty. 2003. *Death of a Discipline*. New York: Columbia University Press.

Tee, Kim Tong. 2003a. "Minor Literature, Polysystem: The Problem of Language and the Meaning of Sinophone Southeast Asian Literature." In *Contemporary Literature and the Ecology of the Humanities* (*Dangdai wenxue yu renwen shengtai*), edited by Wu Yaozong, 313–27. Taipei: Wanjuanlou chubanshe.

Tee, Kim Tong. 2003b. Nanyang Discourse: Sinophone Malaysian Literature and Cultural Identity (Nanyang lunshu: Mahua wenxue yu wenhua shuxing). Taipei: Ryefield.

Tierney, Robert Thomas. 2010. *Tropics of Savagery: The Culture of Japanese Empire in Comparative Frame*. Berkeley: University of California Press.

Veracini, Lorenzo. 2010. *Settler Colonialism: A Theoretical Overview*. London: Palgrave Macmillan.

Wang, David Der-wei. 2001. "Bad Boy Ng Kim Chew: Ng Kim Chew's Chinese Malaysian Discourse and Narrative" ("Huai haizi Huang Jinshu: Huang Jingshu de Mahua lunshu yu xushu"). In *Dari Pulau Ke Pulau*, by Ng Kim-chew, 11–35. Taipei: Ryefield.

Zhang, Longxi. 1992. "Western Theory and Chinese Reality." *Critical Inquiry* 19, no. 1: 105–30.

Zhang, Longxi. 1993. "Out of the Cultural Ghetto: Theory, Politics, and the Study of Chinese Literature." *Modern China* 19, no. 1: 71–101.

The Accumulation of Difference and the Logic of Area

Gavin Walker

> The adjustment of the accumulation of men to that of capital, the articulation of the growth of human groups onto the expansion of the productive forces and the differential reallocation of profit, were rendered possible by the exercise of biopower in its multiple forms and processes.
> —Michel Foucault, *Histoire de la sexualité I: La volonté de savoir*

The Apparatus of Capture: Enclosure into Difference

As is well known, Foucault's later work, particularly in the wake of *Discipline and Punish*, is devoted to the great historical transformation in the operation of the control of the human being, the question of how power operates in and through given situations. As he clearly demonstrates, the transition to the modern capitalist order, centered on the network of commensurable nation-states, inaugurated a historical transition from the "anatomo-politics" of the control of the human body itself (characteristic of the feudal order) to a form of "biopolitics," that is, the control of the human through the form of "population," a means for power to deploy itself through "the entire political technology of life." Foucault's analysis of the manner in which this order continuously reproduces a mechanism that ensures its own functioning by positing the form of population, which the order then utilizes by retrospec-

positions 27:1   DOI 10.1215/10679847-7251819

tively extending itself on the basis of this form, as if it had been there all along, is deeply connected to Marx's analysis of the historical presuppositions for the becoming of capital. Although Foucault takes up the question of territory, or territoriality, as an essential subjective technology in the management of the population, we ought to add to this analysis the question of "area." Taken in this sense, area differs from territory in a number of decisive ways: territory connotes a spatial register that is bounded and countable, that has definite boundaries and limits, and that functions as a site of location, a site *in which* something arises. That is, something is located within a territory—territory names the surrounding configuration in which something occurs, and is directly related to the problem of land in political economy. The control of territory is a question first and foremost of relations of property, it is the control of an inherently limited resource, a resource that can be bought and sold as a commodity but cannot be produced in the capitalist production cycle as a commodity. This aspect of territory is the essence of its centrality for the state, for organs of control over human movement, and for accumulation itself. But "area" is something slightly different. Naoki Sakai (2009) has given us the most critical schematic definition of this term: "When the space demarcated as interiority is determined as enclosure of the national language, national culture, and national subjectivity in the tropics of translation, I call it 'area.'"[1]

Area is typically understood as a form of territory, but area is not merely a zone or domain in which something occurs. Rather, area connotes the basic unit around which the form of *belonging*—the interior as enclosure—is configured; that is, area is not a location per se, but rather indicates the subjective technology of *localization*, the generative flow according to which certain objects are differentiated from each other, and made to belong to an order that supposedly "explains" their existence. In other words, area is a technology that intervenes in the original flux of bodies, words, affects, practices, and so forth, and allows them to be overcoded with an ordering mechanism that renders them "explainable." Area in this sense is perhaps the quintessential modern subjective technology par excellence, in the sense that it is the force that grounds the reproduction of the order itself, by grounding the order's own device—population—with a clearly defined template for the formation of putatively "natural" forms of hierarchy. Without this technology, the form of belonging—national, cultural, linguistic, religious, and so forth—cannot function in the way that we have come to consider stable. That is, the modern world we inhabit is ordered according to this form of area—it is ordered in such a way as to make area and the forms of belonging corresponding to it the central and unassailable substratum of categorization, hierarchy, and relation.

It has been remarked that Foucault's analysis of the term *population*, although filled with rich insights, does not necessarily draw a clear theoretical relation between the role of population as an ordering mechanism and the ways in which biopower, as a historical stage of power's deployment, operates; nor does it extensively develop the political consequences of this relation.[2] That is, why does the operation of biopolitics in contemporary capitalism specifically necessitate the form of population? Biopower, the investment of life into governance, is a sequence of operations, functions, and intensities that delegates into various existing spaces the control and calibration of the smooth functioning of domination. This occurs not strictly through dispossession, oppression, and so forth, but rather through the formation of investments, the direction of desires and affective sensibilities, the deployment and management of operations of identification, continuity, and stability. That is, biopower can be distinguished precisely by its tendency of insinuation, its effect of managing its own outside by conceiving of and deploying assemblages capable of regulating even those things that are strictly speaking not "ready at hand" for it. What links Foucault's analysis of population in this sense to the problem of area, is that these terms are intimately related to what Marx referred to in his analysis of the "so-called primitive accumulation" in the last eight chapters of the first volume of *Capital*, that is, to the problem not only of how the modern world order reproduces itself but also how this order relates to and returns to its own origin.

Today, despite the wide variety of perspectives on the question of primitive accumulation and the return of the enclosures as an analytical site, the general tendency in such strategies is often to assume that enclosure means the divesting of the "native" population of their difference, and the submission of this dispossessed population to the imperial "One," to the flattening homogeneity of global capital and imperial nationalism. But there are numerous theoretical and political problems with this modality of understanding. What is at stake in the question of enclosure is, in my view, something more fundamental for historical inquiry, for examining the historicity of the great mass of social practices that constitute a given moment or set of conditions. Enclosure does not simply mean the submission of difference to the same. In fact, we might profitably understand it as precisely the opposite: enclosure does not mean enclosure *from* difference, but enclosure *into* difference. Enclosure indicates the formation of frameworks for commensurability, which is not at all the same thing as "the Same," understood as that which excludes difference or differentiation. Rather, commensurability means that, on some level of the composition of elements that constitute two or more sequences, an encounter can take place. That is, two groups of effects can encounter each other as *capable of this* encounter, as mutually countable within the range of effects that each sequence gathers. In my view, and it is this point that I would like to emphasize above all others, the process of enclosure should be understood not merely as the unilateral act of dispossession and expropriation, but as an immense historical tidal wave of creation that sweeps through the social field toward the economic, political, and juridical constitution of a network of commensurabilities, a process of making-commensurable, and an installation of becoming-commensurable at the heart of the general economy of society.³

But how does this enclosure as a creation of the new emerge? What institutions carry out the acts of enclosure that enable the new system to overcome the historical boundaries thrown up by the ancien régime? The chaotic and unordered flux of elements prior to the origin of the modern order is overlaid with a grid, a finely calibrated latticework of computation whereby this flux will thereafter become specific difference—this grid effect is exactly what Deleuze and Guattari call the "apparatus of capture."⁴ In a particularly important theoretical move, they utilize, like Marx, precisely the example of racism as the most exemplary site of analysis for the question of the enclosures. Capture, or enclosure, is constituted by two operations: "direct comparison and monopolistic appropriation. And the comparison always presupposes the appropriation. . . . The apparatus of capture constitutes a *general* space of comparison and a *mobile* center of appropriation," "a white wall/black hole system," or regime of "faciality [*visagéité*]" (Deleuze and Guattari 1980: 555; Deleuze and Guattari 1987: 444–45). What they mean by this "white wall/black hole system" is exactly the question of commensurability. That is, in this system of capture, all differences are placed onto a white wall, the expansive and broad social plane in which all differentiations are inscribed on the surface. Here, the system of apparent social contradictions is installed by the advent of capitalist relations of production, and not something that preexists it. Capitalism is a social system in which the forms of differentiation under it do not preexist its origin, they do not furnish the "depths" of the truth of its systematic nature. Rather, it is a system that is organized to enclose the existing flux into a planar field of signification (the white wall, in which all differences are lined up) and sites of subjectivation (the black holes), wherein desire itself is put to work for its own ends. Thereafter, the elements that have been enclosed or captured will furnish the putatively "natural" grounding of the system's capacity for thriving on its own apparent contradictions.

This formulation of the "apparatus of capture" is essential to understanding why capitalism requires a process of enclosures, the question of how and why a logic of area is required for the maintenance of the modern world order. Economically speaking, the process of primitive accumulation and its enclosures is generally understood through the historical deterritorialization of the peasantry from the agricultural community producing for self-sufficiency and its reterritorialization in the industrial form, thereafter subordinated to fixed capital. But what Deleuze and Guattari point out in this question of faciality is that enclosure, while a moment of appropriation, is also the formation of a preorganized system of positions, a creation of boundaries within which comparison will thereafter take place, strengthening the effect of capture every time it is undertaken. Enclosure means that there is no exclusion. In other words, enclosure or capture operates by forming a social field in which all differences are accounted for in its own range, in which all differences are placed onto the white wall, interspersed with black holes of passion. Thereafter, there is no such thing as the substantial "exterior," the absolute exterior. Instead, there come to be "internal exteriors," pockets of energy in the process of forming subjects that show that this white wall/black hole system is a vast expanse of gaps. Obviously, this understanding of enclosure is intimately connected to the history of racism. It is by now a relatively common understanding that racism is not, and has never been, an exceptional or excessive practice to the formation of the modern state system, but rather one of its essential founding gestures. It is this originary form of racism that can be understood precisely through the term *enclosure* and the way it functions. In other words, "the state of actual domination—and the production of the majority that corresponds to it—takes the form of the constitution of a face. Or, more precisely, what is at stake is the *unity* of a face." This enclosure within specific difference is the "producing of a model of identity and normality in relation to which deviations can subsequently be detected." Racism, as an act of enclosure in this sense, is not merely a "strategy of exclusion from without," but a "strategy of constitution from within" (Marrati 2001: 208–10). It is exactly this element of constitution that is the most critical aspect of the process of primitive accumulation. If we merely draw attention to the element of exclusion, we will always miss the truly astonishing biopolitical violence of enclosure and the operation of the capitalist accumulation cycle, in which the violence of maintenance, care, and sustaining is always being covered over. This element of enclosure is critical on the level of historiography precisely because it can be an instructive direction from which to rethink the questions of "late development" and the relation between the general and the specific that are implied by the problem of area, between the logic of capital and the logic of the concrete situation. It is precisely in the explication of this space of tension that Uno Kōzō brilliantly summarized this problem by clearly showing us how capital overcomes its own basic limitation:

> Capital is something that cannot directly produce the labor power commodity, but through the formation of the relative surplus population in its accumulation process, *it can indirectly produce it* (my emphasis), so to speak [*iwaba kansetsuteki ni seisan suru*]. Thus capital gains a method of releasing the supply of labor power from the limitations of the given natural population: the development of what Marx referred to as the special character of the law of population in capitalism. This is precisely the fundamental basis through which capitalism makes itself into a specific form of society. This is why the "historical and moral factors" contained in the determination of the value of labor power must be formed in such a way as to be commensurable with capitalist production. (Uno 1973b: 132)

That is, to control and maintain something that it in fact cannot control, capitalism forms a means of producing the labor power commodity "as if" it were, in fact, under its direct jurisdiction. What it requires is a form of delegation that can formulate social-historical institutions capable of inciting instances of these "historical" and "moral" aspects (those aspects organized around the technology of substantialization called "area") of the field of life (from which labor power is drawn) that are "suitable" for capitalism's own reproduction. Thus, capitalism's specific form of population is a complex aggregate of techniques that are overlaid like a grid on the existing "natural" stratum of bodies, words, physiognomies, affects, desires, and so forth, that recalibrates and reformulates them as "countable" or "computable" by capital as inputs for its circuit process. This "countable" or "digestable" aspect is precisely what results from capital's need to "fold" back this process into itself, to posit the external limits imposed on it as if they were imposed from within itself. Therefore, what Uno extensively theorized in terms of capitalism's unique "law of population" is precisely the *katechonic* moment that "allows" the exterior into the interior,[5] the mechanism or assemblage by which the excess of subjectivation that is produced can be incorporated into the smooth circuit of accumulation:

> Through the law of population, capitalism comes into possession of *a mechanism which allows this excess of the commodification of labor power to pass through it* [*rōdōryoku shōhinka no muri o tōsu kikō*]. This is precisely the point on which capitalism historically forms itself into a determinate form of society and, further, is what makes it independent in pure economic terms. Like land, this is a so-called given for capitalism, one that is given from its exterior, but unlike land it can be reproduced, and by means of this reproduction becomes capable of responding to the demands of capital put forward through the specific phenomenon of capitalism called crisis. (Uno 1973c: 426–27)

Capitalism itself does not produce labor power, but rather produces assemblages or mechanisms (*kikō*) that "transmit" or "allow through" (*tōsu*) the effect of its excessive or irrational moment (*muri*), this folding back into itself. It is no accident that the term Foucault uses to describe how power directs, manages, and organizes the specific form of population is exactly the same as this "transmission": it "guides" or "conducts" (*conduire*) the formation of the "historical" and "moral" elements of social life by means of this form called "population."[6] The political technology of area, and indeed the form of area studies, which I will address at length later, that is, the *schema of civilizational difference* according to which belonging is naturalized and substantialized in terms of a network of areas, is a decisive technology for the "historical" and "moral" formation of the border, the "historical" and "moral" elements that capitalism requires the formation of in order to then use their supposed presence to engender itself from them. It is no accident that Jean-Luc Nancy traces the formation of the idea "West" (the essential expression or archetype of "area") to this logic of a structure that paradoxically presupposes for itself precisely those effects it organizes: the "very Idea of the West" appropriates its own origin precisely so that it can identify itself "around its own pronouncement (*profération*)," compulsively returning "to its own sources in order to re-engender itself from them as the very destiny of humanity" (Nancy 1983: 117–18; 1991: 46; translation modified). Indeed, capital's sustained reproduction, and by extension the reproduction of the schema of the world on the basis of civilizational difference, depends on this border—or more precisely, its formation through the subjective technology of area—to release itself from the limits that it creates for itself. Area is the decisive technology that intervenes to propel forward the "adjustment of the accumulation of men to the accumulation of capital" that Foucault brilliantly analyzes. The investigation of area in this sense clearly shows us that the form of national belonging, the form of belonging itself, is not a question of control that merely parallels the question of capital. It shows us, rather, that because capital must always from the very outset become involved with its "outside," involved with the "historical and moral factors" that allow for a supply of labor power through the formation of the relative surplus population, the form of belonging is something *through which* capital accomplishes its project. Among many other derivations from this point, it is thus clear that capital can never be challenged through the valorization of "inherent" or "stable" qualities of belonging; this stability is instead a "semblance" (*Schein*),[7] a schema deployed by capital itself. Capital deploys this putative stability as a stability, but can never succeed in erasing or eliding its "putative" character. Because capital cannot itself reproduce the energy that drives on its own functioning (labor power that can be commodified), capital must always act *as if* this stability is present. Hence, when capital delegates functions to its outside—the form of area—for the reproduction of these elements that it itself cannot produce, it is always receiving back a supply of something irreparably under the condition of this "as if." In other words, the form of belonging that emerges in conjunction with the technology of area is not itself a stable form that then balances capital's equilibrium. Rather, capital is always wagering on the role of the "historical and moral elements" contained within this form of area to maintain this form of belonging *as if* it were stable. This is why area is such a decisive question: capital is always inserting into itself and relying on something extremely volatile, something that is always "on the edge of the void," something that is always in danger of imploding under the weight of its fiction. Hence, for Foucault,

> Biopower was without question an indispensable element in the development of capitalism; the latter would not have been possible without the controlled insertion of bodies into the apparatus of population and the adjustment of the phenomena of population to economic processes. But this was not all it required; it also needed the growth of both these factors, their reinforcement as well as their availability and docility; it had to have methods of power capable of optimizing forces, aptitudes, and life in general without at the same time making them more difficult to govern [*assujettir*]. (Foucault 1976: 185)

The most important of these "methods of power" according to which the "accumulation of men" could be calibrated to and complimentary with the accumulation of capital, is the form called "area." Without this mechanism that allows the form of belonging to be a direct factor of production, capitalism could not have emerged as a total social system. Hence, we must theorize the question of area, not only in terms of the operations of biopower pointed out by Foucault but also in terms of the process of the primitive accumulation of capital. It is, in fact, one and the same process, the process of enclosing the flux of bodies, words, statements, histories, and other purely heterogeneous sequences of meaning *into* the units of *difference*. Foucault's work (and the economic correlates to his analysis that can be found in the work of Marxian economists like Uno) shows us precisely that the structuring of power through the articulation process between capital's accumulation cycle and the reproduction of the state (through its utilization of the image "nation") can never be criticized from a standpoint that merely valorizes the distribution and apportionment of a specific difference characteristic of the modern international state system. Rather, the decisive point is that such schematics, because they in effect take this apparent "split" between the general movement (capital's self-deployment) and the particular gradient of its localization (the putative "cultural" substratum) at face value, cannot get us out of the historiographical reinforcement of the binary structure of "the West and the Rest," but rather remain securely within this schematic. These schematics tend to imply that "cultural specificity" is the site of potential, the place from which capital's smooth logic is punctuated and interrupted. But capital, or indeed "area," is not an object that exists in its full plenitude somewhere. Certainly, capital phenomenally appears whenever surplus value is computed and redeployed as valorized. But both capital and area are in essence social relations rather than concrete objects. In other words, capital names a systematic network in which a massive sequence of social practices are aggregated and linked. Capital is a relation that emerges from within the field of practices as a whole and recodes their ordering mechanisms, placing terms into relations that did not previously obtain. As a consequence, capital ensures that by coding vast fields of practices with its own internal directionality, practices which are apparently "outside" of its logical operation are neither troubling nor disruptive of its function. Rather, as Marx and many later commentators have emphasized, the trick or "magic" of capital as a social relation is that it always acts "as if" it emerged into the world perfectly formed, "as if" the moment of enclosure was not its founding gesture. Because capital has already enclosed the earth, we cannot disrupt capital's smooth function by imagining the possibility of de-enclosure in the strict sense. Instead, as Marx emphasized, capital can only be confronted on a level of immanence and displacement, from within the spaces opened up by its own logic.

Enclosure is not the movement of subjecting specificities—language, culture, nation, ethnicity, and so forth—to the developmental tendencies of generality. It is the movement by which simple and pure heterogeneity is violently formed into a specificity, into a category of difference, a movement of capture that "is not merely a question of foreclosing freedoms by divesting labor of property or means of production . . . it is also the work of inciting individual desires in ways that are useful for capital, not to mention critical in its relentless drive for self-expansion" (Gidwani 2008: 197).[8] Enclosure names the aggregate of acts that generate specificities. Enclosure in essence is the conceptual border zone wherein one set of historical boundaries has been breached and another arises, the limit-space in which the process of primitive accumulation's incessant and cyclical need to begin again is socially exposed and rendered visible. What I mean by "enclosure," and indeed by "area," is defined with a precise conceptual rigor in the work of Carl Schmitt (1974, 2003): it is an originary *act*, an "*Ur-Akt*" that constitutes "the *nomos* by which a tribe, a retinue, or a people becomes settled, that is, by which it becomes historically situated [*geschichtlich verortet*] and turns a part of the earth's surface into the force field [*Kraftfeld*] of a particular order" (1974: 40).[9] That is, this is the process by which a sequence of bodies, words, statements, practices, rituals, and so forth, are exposed to a series of effects that thereafter allow these elements to be computed as a sequence that is sequential in the same manner as a number of other sequences.

This is precisely why it is so crucial to understand enclosure not only for its "phenomenal appearance" whereby it is simply reduced to the violence against difference. It is much more important to understand the much greater violence, the refined violence that is almost impossible for us to grasp, the violence, always subjected to its own erasure, of the very creation of difference itself, and specifically, that strange modern form of the organization and categorization of life called "civilizational difference." This violence is precisely what Foucault identified with the advent of a society of control overseen by the operations of biopower: the violence of care, the violence of sustaining, the violence of maintenance, the violence of inclusion. If we seek to understand the rapidly shifting social logic of the contemporary moment and extend its lessons to our own spaces of practice, the existence of this violence is much more critical for us to understand today than the violence of exclusion is.

In my view, focusing on questions like the form of "enclosure" can allow us to bypass the dead end produced in the encounter between contemporary universalisms and particularisms that tend to obtain in the analysis of area. It shows us clearly that the writing of history as a technique of resistance must not simply valorize the existing systematic structure of the world, formed from the unit of the nation-state. Doing so simply ignores the fact that the nation-state form itself is a formation stemming from capital's impossible origin in enclosure. This is precisely why attempts to overcome or evade the problems of the inherited organization of area studies by writing "resistant" histories from the periphery or "nonhegemonic" area can never escape the general economy of civilizational difference. Simply inverting the logic or directionality of the distribution of positions in the area studies modality cannot disturb its structure: in this sense, supposedly resistant histories that remain in the economy of area itself rather parasitically depend on and feed off of the more "traditional" area studies. Without understanding the question of enclosure, the originary or primal movement of history that generates the schematic or grid of civilizational difference as an order itself, we cannot produce a supposedly resistant history. In fact, *enclosure* here means something extremely broad, something that we might also call "bordering" or the "regime of translation," or indeed simply the formation of area. Primitive accumulation is an essential question for the problem of area: this analysis demonstrates that area itself is not a static process but a generative flow, an originary social-historical overtone that gathers into itself the ongoing reverberations of the original appropriation and division of the modern world, an endless echo that grounds the fiction of the substantiality of civilizational difference.

We can clearly see that the subjective technology called area deploys itself in a complex and dense parallax with the expansion of the accumulation cycle. Area in this sense is something like the "spiritual concrete" (*geistig Konkretes*), formed on the level of theory in the accumulation of differences into an international world. This spiritual concrete, the concentrated material force that expresses the historical sedimentation of the functioning of this subjective technology, demands an approach to its actuality, its contemporary level of operation. What should we make of the form of area studies, the broad discipline in which this force is taken as the given and substantial starting point, when we approach it from this rather abstract sense of the term *area*?

The Logic of Area and the Ethics of Area Studies

Today, in a world characterized by the dispersion of the concentration of the productive forces, an increasingly multinational composition of global finance capital and its specialized class of handlers, it is relatively common to hear that the problems of area studies and its critique are no longer relevant. This argument tends to be made as follows: area studies depended on a world characterized by the classic mid-twentieth century structures of alignment: the US-aligned world, the USSR-aligned world, the so-called nonaligned world, and so on. But, so this logic goes, today the world that is implied by this organizational schema itself no longer exists, and therefore the problem of area studies has ceased to be an essential target: it is a "remnant" that is "withering away." But I want to argue here that it is in fact exactly the opposite, that we will miss something crucial in the question of area studies if we imagine that it is no longer a problem for thought and politics simply because of the process of "globalization." In fact, paradoxically, it is the current moment of the integration of the world in which the problem of area studies becomes most decisive.

It is well known that the global Cold War system was the basis of the formation of area studies as we know it today, although not the starting point or origin of the drive that lies behind this disciplinary formation. Throughout the Cold War, the political alignments of area studies depended largely on the schema of the enemy, and knowledge of discrete units of territory that would be characterized by their own native knowledges was an essential element of the preemptive expansionism practiced in this dyadic schema of the world. The areas or fields that were observed, studied, and catalogued functioned as the building blocks of knowledge used to construct a set of representations of the world, a set of representations that buttressed the logic of the dyadic globe: there were alliances, which were by and large putatively "civilizational" in nature (based on the prior existence of the discourse of the West and the Rest), and then there were the "areas" that contained peoples who may or may not be amenable to Cold War alignment. This knowledge gathered, therefore, was directly in relationship with US imperialist policy. It was a wing of policy, a movement of the process of creating alignments, destabilizing hostile alignments, and so forth. The specific knowledge of the "other" space was essential because unstudied spaces that could not be categorized were unacceptable within the field of macrolevel power relations. Therefore at its height, let us say that area studies was the form of epistemology corresponding to the organized social forces of the Cold War system. In a simplistic and obvious sense, we can say that since these organized social forces have mutated, changing their form and direction, area studies as the epistemic field corresponding to them must also have changed.

But this crucially ignores another element, what I will call the "ethics" of area studies. If we accept that the form of the world today differs from its appearance at the height of the Cold War system, that its power relations, level of social forces in the global conjuncture, and forms of knowledge circulation are radically different, then why is area studies even more of a problem now, as I claimed I would argue? Investigating this question shows us something important about the sense of area as a subjective technology, as I outlined in the first part of this essay: it shows us something important about the immateriality of area or "civilizational difference." The world today is unquestionably one in which the diffusion and multipolarity of the operation of power demonstrates its vast difference from the bipolar geopolitics of the Cold War. But the problem is that area studies knowledge is still gathered, acquired, and utilized as if this former world continued to exist. Given this radical split between the organized social forces underpinning the total systematic situation and this untethered form of knowledge production and knowledge articulation, we must ask, if area studies no longer serves to effectively "explain" and thereby articulate knowledge in the direct service of imperialism, why does it continue to exist in the form of knowledge for knowledge's sake? Why would anybody be interested in this interpretive sieve for the consumption of information? What forces and relations sustain such a mode of inquiry?

This is where the ethics of area studies becomes an essential problem for us. The classical developmentalist model of area studies is one that is structured on the form of the nation-state as given and on the form of the national people as a unit commensurable with other such units in discrete spaces; in a perfect cyclical recursion, the "area" in which the "people" are "discovered" is the lever that explains their practices, statements, significations, and so on, yet it is the putatively prior existence of the national people that is in turn mobilized to "explain" the form of the "area," the institutional significations of the state-form, and so on. Thus area studies operates in a logical circuit that always refers back to itself, short-circuiting the dilemma of the articulation between the nation-form and state-form with the rhetoric of the given and identitary. As is well known and obvious, this explains nothing whatsoever, yet it remains today the decisive popular mode of "explanation" for the alterity of the other: the other is other in relation to me precisely because something about their practices "others" them, and the origin of this something can be found in the other itself, whose very existence supposedly furnishes the "proof" of their "alterity," precisely because the other is part of a commensurable sequence of those who are other, yet comparable, those who "belong" to putatively different areas, but who can encounter each other *as different* along the lines of the model of communication, and so on ad infinitum.

Why today, when the forms of social organization and global modes of life practice are undermining the possibility of employing this type of exercise in absurdity, would this remain such an important apparatus for the dominant sociality? It is a site in which we can see this problem of ethicality: the fact that there is no necessity for area studies. Now that area studies cannot point to a total social system in which it plays a decisive axial role, we have to examine the problem of area studies as one also of the direction of desire or investment/cathexis (*Besetzung*),[10] as one characterized by a specific ethics and specific drives. Area studies is something for which there is no exigency, but for which there is a longing. It is an identitarian *obsession*, a means of comfort. As a strategy, it is primarily concerned with insuring that the position of the self remain unassailed and stabilized, and therefore must make the other both an absolute difference and an absolute commensurability at the same time: the other must be the index of all that the observer is not, and at the same time, the grounding or naturalization of the other's alterity reveals that the other is commensurable, can be related to, but only in the form of a hierarchy of knowledge, only in the form of a "communication" across "civilizational difference." This desiring positionality of area studies is outlined in Nietzsche's (1980: 265) early critique of historicism: "All that is small and limited, rotten and obsolete, gains a worth and inviolability of its own from the conservative and reverent soul of the antiquated person who migrates into it, and builds for himself a homely nest [*ein heimisches Nest*] there."

Area studies in its classical modality appears increasingly frantic and panicked, because without the Cold War system, its foundation as a specifically ethical field, a decisional space, is revealed: to practice area studies in the classical mode now has an exhibitionist dimension, a performative narcissism which cannot be explained except to say that it is an ethic, a choice to engage in this modality of operation, because there is no longer an ulterior motive. The motive and drive of area studies does not need to be sought in its depths, in its secrets: the drive is inscribed on the surface, it is exposed, obvious, exhibitionist in form and content. As I have stated, area studies is a form of knowledge that corresponds to another correspondence: that between the US Cold War military apparatus and the hegemonic forms of US knowledge production. But now, these two directions have largely split apart. The site of area studies, in other words the university and the former "areas" themselves (that is, the "Rest" in the West and Rest dyad) are now bifurcated and made transverse by an increasingly generalized social knowledge, bypassing area studies itself, a disciplinary apparatus now reduced to the status of a useless and debilitated filter that has been exceeded by lived historical actuality. As a consequence, the accumulated knowledge that is ceaselessly regurgitated within the last remnants of area studies appears as a final line of defense for a "national" knowledge: in this way, area studies *increasingly*, not *decreasingly*, resembles a colonial form of knowledge control. It is not that the further the material basis of the colonial systems recede into the past, the less colonial the knowledge will be: area studies today exhibits the precise inverse.

Area studies is now a directly colonial division of power relations, whereas before it was merely in gestation: this corresponds to the problem of the colonial system and the postcolonial as a condition. During the existence of the Cold War system, area studies could be understood as a field of epistemology devoted to the classifications and schematics, the blueprints needed for a broad-based colonial form of disciplinary governance of spheres of influence. In the aftermath of that system, area studies today appears not as a disciplinary form, but as a politics of control, devoted to not giving up on the schema of the world developed through the direct colonial governance of large swaths of the world. That is, "despite the advances of the critique of Orientalism in the 70s, the critique of the (latent colonial structure of knowledge) in cultural anthropology in the 80s, and the development of postcolonial studies in the 90s, area studies related to East Asia have still not arrived at the stage of repudiating the system bequeathed by the Cold War period . . . the system of area studies itself is no longer capable of its own self-justification" (Sakai 2008: 215). Nietzsche (1980: 330) famously drew our attention to the two tendencies of the operation of history to comfort itself in the face of its demand for life: the "unhistorical [*das Unhistorische*]," in other words, the "art and power of forgetting, and of drawing a limited horizon around one's self," and the "superhistorical [*das Überhistorische*]," the "power which diverts the eyes from becoming to that which gives being an eternal and stable [*Gleichbedeutenden*] character." Area studies today is devoted to precisely these historicist mobilizations: the limiting of analysis to an outmoded horizon of thought, the eternalization and stabilization of the civilizational difference, the schema of area that allows for a stable mapping of belonging. If area studies is to be reorganized, the decisive point is not simply that it should cease to operate through the givenness of certain areas as the explanans of development, but rather that it must return to its own origins to examine the forces at play in its own constitution. In this sense, the possibilities posed by the crisis of area studies, the exposure of the decisional space in which it operates today with no necessity, without a coherent referent, also allows us an opening to consider a new ethicality, a new practice of life toward which this crisis can lead us. In other words, it is a situation in which we directly face the decisive Nietzschean moment: "History [*Historie*] *must* itself resolve the problem of history. Knowledge *must* turn its barbs [*Stachel*] against itself" (306). That is, area studies is not a discrete field, a certain "particular" problem. Rather, it is one organized moment of the general question of the epistemic hierarchy according to which the "world" itself is organized. Area studies as a discursive formation cannot be "overcome" by simply arguing that it has run its course or exhausted itself as an explanatory force; rather, what has potential in the crisis of area studies is this tension itself, "this impossible 'no' to a structure that one critiques yet inhabits intimately," which is "the deconstructive position, of which postcoloniality is a historical case" (Spivak 1999: 191).

Instead, the only way to bisect the assumptions underlying area studies is to create anew a form of the sociality of knowledge capable of responding to this coloniality of power: such a sociality must from the very outset give a new meaning to the question of complicity, perhaps, in fact, to the question of shame. In a well-known series of letters, Marx writes as follows:

> I assure you, even if one has no feeling of national pride at all, neverthe-less one has a feeling of national shame. . . . This is a truth which, at least, teaches us to recognize the emptiness of our patriotism and the abnormity of our state system, and makes us hide our faces in shame. You look at me with a smile and ask: What is gained by that? No revolution is made out of shame. I reply: shame is already revolution of a kind. (1975b: 133–34).[11]

It could be suggested by a certain logic that shame is merely the obverse of pride, and therefore remains tied to the same pivotal logic of the national. But this is not the case if we add the essential third pole of this relation, which is complicity. Today, in the era of the eclipse of the classical sover-eignty of the form of the nation-state, shame could perhaps be termed an anachronistic site of analysis—but the question of shame, the operation of shame is an essential moment for us today, not one of retreat into national forms of subjectivation, but rather a projection for the possibility of a new recombination, a new sociality of life to come. The time of shame is typically understood as the past, as a topos of articulation bisected by the flows of the past, but in fact the arrival of shame is always something immediate, some-thing that is neither past nor futurally deferred but a question of the history of the present, because it is an ethics, a practice, and not a preexisting object. It is never a "code which would tell us how to act," but a site or articulatory location in which can be posed the question of the "relationship you have to yourself when you act" (Foucault 1996: 380). Tazaki Hideaki (2007: 115) has incisively formulated this possibility in his own examination of this state-ment: "The feeling of shame is, as Marx suggests, anger directed toward my 'self,' toward my own existence. The object of this anger is not simply 'American,' 'Japanese,' and so on. To put it another way, it is not merely the *object* of identification but a turning of this anger toward the *operation* of this identification itself." Through this turn toward the self, we have to pursue a different strategy, one of refusal and removal, one of the subtraction of the basic element of area studies as a circuit of reproduction: ourselves. Area studies is a kind of policing agency, a disposition, an ethics, a decision. It is held in place by requiring collusion, by tapping into our complicity with ourselves, by accessing the fact that we cannot wish away the residues of the "national shame" that Marx refers to.

But it is precisely because area studies relies on the deepening of this complicity into acquiescence, into an ethics of the naturalization and substantialization of contingent hierarchies, that it sets up a situation in which it can no longer justify itself. Area studies functions by emphasizing and highlighting the moment of vertical relations of domination in this topos of complicity. But complicity (the affective response to which is shame) itself also contains a potential, a virtuality, which is the horizontal connectivity of complicity itself as a location: no one can be altogether separate from the forms of capture that exist, since "shame is grounded in our being's incapacity to move away and break from itself" (Agamben 2002: 105–07). To the extent that all people share at least one location, which is the irreducible proximity/distance from themselves, from the "I," the acknowledgement of complicity itself—for instance, national shame—can be a recombinant function, a zone of flux for us to attempt forms of encounter with each other diagonal to the mechanisms of capture.[12]

Such a possibility of recombination can lead us to a new sense of contemporaneity, a new sense of the materiality of social relations produced through such a radical withdrawal. As Giorgio Agamben (2008: 16) has pointed out in a recent essay, "contemporaneity is a singular relation to one's own time, which adheres to it and simultaneously distances itself." He continues, "The contemporary is rare. To be contemporary is first, a question of courage: because it signifies being capable not only of keeping our gaze fixed on the gloom of our era but also to perceive within this gloom a light which is directed towards us, which is infinitely distanced from us."[13] This courage stems from the possibility of shame, the irreducible distance in time between the self and the "I." Thus to be contemporary is not to encounter the other in the form of the communicative event, the "dialogue"—rather, it is to encounter oneself as other, and to encounter other singular human beings through a vicarious relation, that is through their own encounters with themselves:

> The time will come in which people wisely refrain from all constructions of the World Process or even of history of man, a time in which people in general no longer consider the masses but once again think about individuals who construct a sort of bridge over the chaotic stream of becoming [*den wüsten Strom des Werdens*]. These people do not set out some sort of process, but live timelessly-contemporaneously [*zeitlos-gleichzeitig*], thanks to history [*Geschichte*], which permits such a combination. (Nietzsche 1980: 317)

Thus, each of us "must organize the chaos in oneself by 'thinking oneself back' to one's own true needs" (333). This contemporaneity is not a given, but a project—what Foucault described as the "history of the present," what Uno described as the rare and intense potential of the "analysis of the contemporary situation [*genjō bunseki*]."

This is the materialist possibility pointed to by the problem of national shame—shame directs us toward the self, through an internally directed anger at the mechanisms of identification. But we cannot escape by simply wishing these mechanisms away—rather, we can find an exodus or removal from the logic of sociality based on the nation-state by deepening this national shame into a concrete historical-material practice, by attempting "to live timelessly-contemporaneously," rather than consoling ourselves with the circular logic of the area. Thus "when I relate myself to myself as if to something which is directly another, then my relationship is a material one" (Marx 1975a: 53). It is a materialist sociality that finds its lines of connection in the distance between self and self, and in turn, discovers that this spacing between the self and the I is a topos where others too overlap. But this flickering, fluctuating site is only accessed by means of an ethic, a practice of knowledge that holds the self at its center. Marx directly articulates the sociality of this site: "When I am active in knowledge, etc.—an activity which I can seldom perform in direct community with others—then my activity is social, because I perform it as a human. Not only is the material of my activity given to me as a social product (as is even the language in which the thinker is active): my own existence is social activity, and therefore that which I make of myself, I make of myself for society and with the consciousness of myself as a social being" (Marx 1975c: 298). In his examination of Nietzsche's affirmative grasp of life against the deadened disciplinary field of received history, Minatomichi Takashi (1998: 106) has argued that "the contract/promise with the humanity to come is first and foremost a contract/promise with the self itself: the self as interlocutor who introduces the self is the first other." At such a moment of self-encounter the "we" that is so heavily relied on "loses its identity as a subject, perhaps for example, forgetting the boundaries of the 'nation,' no longer reducible to the 'national people,' and going beyond the 'good European,' but instead multiplying its linkages to completely different, entirely multiple others" (110).

Without examining in light of a new form of direct, material-historical practice the epistemic hierarchy and logic of the "original appropriation" on which "area" is founded, it will continue to operate unchanged. Creating a sociality that can exceed area and sustain itself through its own constant recreation requires a shift from the fantasy of the subject of history to the possibilities of recombination. Just like the renewal and constant expansion of the aggregate capital, area studies as a form is incapable of sustaining itself without a connection to the material flow of social life. Precisely in this sense, area studies thrives on the complicity it produces. But in doing so, in making complicity the essential raw material of its reproduction, area studies constantly generates the potential for its own obliteration: turning our attention not onto the terms of that complicity, but onto that relational space of complicity itself can lead us to develop a new sociality, a new type of recombination capable of responding to and being responsible for, the contemporary coloniality of knowledge, and its fundamental relation to every space in which the "I" can come to be. Franco Berardi has argued that

if we wish to define "what is to be done" for our time, we must concentrate our attention on the social function of cognitive labor. This would not be to construct a subjective vanguard that functions as a collective intellectual, but to find epistemic, communicative, and political apparatuses [*dispositivi*] that provide cognitive workers the possibility of acting as transformers [*fattore di trasformazione*] of the entire cycle of social labor. The problem of our time is the creation of a recombinant function, a function of subjectivity able to cross the various fields of social production, and to redefine its contents in a paradigmatic frame that is not dependent on profit, but on social utility. Intellectual labor is no longer a social function separate from labor in general [*lavoro complessivo*], but has instead become a transversal function within the social process. (2004: 181)

This is not to posit a radical outside today, but a mutually imbricated outside, the overlapping spatiotemporal zones of inside and out, which is, in the final analysis, the space of identification itself. We must emphasize here that the "inescapable" historical present is at base, nothing less than the "inescapable" "I," the space of flux between interior and exteriority that is produced as a result of the fact that I cannot be other than myself, and at the same time can never discover myself as self-identical with this "I."

If the location of the problems of area is the "I," the central nodal zone around which the problematic, the hydraulics or schematics of power flow, what is this question of the "coloniality" of knowledge that is so essential to area studies? This coloniality is the obverse of the inescapability of the "I"—it is the form or logic according to which the encounter with the other is managed, organized, hierarchized, understood, inserted into general social knowledge. The reason that area studies, or more specifically, the logic of area as circular *explanans* of the schema of cultural difference itself—irrespective of where and by whom it is practiced—is a colonial form of knowledge is that it is split from its systematic origins. It is now a pure desire, a desire to retain a world organized by certain specific hierarchical divisions, a desire more than anything to be sure of one's own essential role as observer: in short, area studies itself is a desire to maintain the stability of the discursive formation of the West. The coloniality of knowledge that is the fundamental mode of area studies can only be escaped from by understanding that the first-order appearance of area takes place within the "I" itself, that it cannot be problematized unless it is first turned toward the operation of the self. There is no comforting, consoling outside to be discovered, no fantasy of a certain "people" or "culture," a "proletarian nation" that would be the messianic hinge of the revolution, a "discoverable" outside that can salvage the mapping of the world of nation-states, and thereby give us assurance. If there is an outside, it must be produced by us through a transversal sociality, a recombinant function that must first and foremost be a type of "self-introduction" to the complicity contained in the "I" itself. This could be exactly the "national shame" that Marx mentions, a "shame" that is not the obverse of pride, but "shame" as a practice, an ethics. Such a practice is precisely what can allow us to constitute a revolutionary zone, a "subtractive" zone of exodus toward a new social form.

We cannot produce a new expressly political critique of area studies, because area studies, understood as a drive, as an elemental force of the apparatus of capture, is capable of sublating into itself any "worldview" and grounding it in the hierarchical commensurability of the international world. Instead, we can only demand a new ethics, a new practice of life that itself evades capture not by attempting to overthrow, on the level of the concept, area studies as a methodology and replace it with another. Such a practice (which has been constantly attempted in numerous situations) only reinscribes area as return and recurrence, a fantasy of reversal that reproduces the same logic in practice. Because area studies is a practice, a constantly rejuvenated social body whose logic is renewed in a type of daily plebiscite, a new ethics of globality can only be a movement toward the question (see, e.g., Spivak 2003) of how to recombine in a different form the essential and irreducible moment of the social. Thus the ethical moment of area studies, and the deepening of its coloniality of knowledge production and knowledge relations today, signals the urgent need for a counterethics, an antimemory that can spur our exodus away from capture. But what lines of possibility exist for area studies? How and through what specific practices could area studies invert its constitutive dementia into a site of potential for intellectual renewal and new social possibilities? What organizations of knowledge would underlie such an effort?

In their well-known joint work, Deleuze and Guattari consistently insisted on the importance of the figure of the schizophrenic as an intervention into existing knowledge practices and as an ethico-political articulation of the possibility of a future form of thought actively dissociating itself in practice from its inherited hierarchical organization. I want to emphasize that their deployment of the figure of the schizophrenic has nothing whatsoever to do with the various forms of eclecticism, political narcissism, or reactionary opportunism that certain of their liberalist followers have imagined. When they explicitly argue that those who practice social sciences should "make themselves schizophrenic" they intend to use this term in a very specific manner; that is, they want to draw our attention to the schizophrenic's ability or tendency to "range across fields" without respecting their prior forms of allocation: "People who are operating on the level of social sciences or on the level of politics ought to make themselves schizophrenic. And I'm not speaking of that illusory image of schizophrenics, caught in the grip of a repression which would have us believe that they are 'autistic,' turned inward, on themselves, and so forth. I mean that we should have the schizophrenic's capacity to range across fields" (Guattari 2009: 59).

This "ranging," the deployment of a transversal line or diagonal practice of articulation is already imbued in area studies from the outset, but in a defensive, reactive form. That is, it is exactly here that area studies falls into a program of substantialization and fantasy: what forms and situates the diagonal range of area studies in its traditional inflection is precisely the "area." Hence area specialists are meant to be somewhat like priests, missionaries, or travel agents, conversant and familiar with a whole range of social significations bound together only by the unstable force of area, the terms *Japan*, *Russia*, *West*, and so forth. Of course, it is exactly this hermetic logic of knowledge, that is, the fact that the "area specialist" is understood as a force of mediation facilitating "exchange," "communication," and so on, that keeps area studies in a permanently outmoded conception of the world, a world that has no need for this type of "expertise," a world in which the density and magnitude of the commodity-form's expansion has eliminated the usefulness of such mediated knowledge. And yet, the fact that area studies is essentially a discipline founded on the logic of the fantasy of the West, the fact that the object of area studies is this unstable zone of the hazard, namely the space of the symbolization, substantialization, and fetishization of identifiability, signals that the space of excess that would remain after the closure of the area studies sequence could hold a great deal of potential. In other words, if area studies could invert or displace its everyday logic of practice by truly "becoming area studies," that is, by becoming the study of how the concept "area" was itself constituted, formed, and maintained, there would be a great potential for area studies as a critical site of engagement within the broadening of increasingly socialized knowledge characteristic of the process sometimes called "globalization." That is, it is only at the moment when area studies has lost its historical role that area studies can clearly be seen for what it is: not a set of practices of introducing "other" knowledge to the realm of *humanitas*, but a set of practices for the self-structuring, clarification, and concentration of the schema of civilizational difference. The trajectory of the symbolic order of area studies now points directly to its "secret" object: the area studies specialist him- or herself. It is the abstract drive inhabiting this fantasy that maintains the bordering effect of the regulative idea "area," even after the subjective technology of area, situated between the biopolitical management of the population and the national fold of substantialization, has shifted its mode of operation.

If area studies were to function in a divergent manner, if area studies could be subject to a form of recombination, we would quickly require a rethinking of the practice of area studies in relation to the problematic of translation. That is, the "area" in area studies is the fundamental institutional form corresponding to the regime of translation on which the formation of national language rests. Area studies is a political technique mobilizing and substantializing the position of the translator, imaging and imagining the translator as mediator, as a filter, as a supposedly transparent medium of exchange, transmission, and communication. But as the work of Sakai, for instance, has decisively demonstrated, translation is itself merely an act of articulation in the social topos of incommensurability, not a mediational transposition between two substantial entities. In fact, this "transposition," in which the putative substantiality of the two sides legitimizes their existence as separate, is precisely what retrospectively creates and forms the commensurable. By rerouting the logic and practice of area studies to its own initial moment, area studies could become a crucial site of contemporary knowledge production, rather than the necrotic body of everything "rotten, limited, and obsolete" that it is today. In considering therefore this potential of the location of area studies, so close to the mechanisms of enclosure threatening a new politics of life today, we should emphasize that the danger inherent in the epistemology of area studies also contains a sequence of potentialities.

In this sense, this inversion or affirmative implosion of area studies can only occur if its practices are deployed in a more schizophrenic manner. Only by exploding the relationship to the Oedipus of area studies, namely, the fantastical relationship to the object according to which meaning is attributed and short-circuited at the same time through the identifying grid of area, a configuration schematized around a tripartite structure of putative substantiality (ethnos—national language—territoriality), can we find something interesting and new, and in any case coherent, in it. Area studies, as a "traditional" discipline of knowledge about the "other" space, always attempts to repress and exclude "incoherencies," and therefore continually tends to present itself as a guardian of coherence—a coherence based on the authority of inherited knowledge, the defense of the existing division of labor, the accepted rationale of positioning within its field, and so on. But, in fact, area studies can only mobilize its rationality, its inherited logic, on the basis of an incredible, stupefying incoherence: the notion that the substantial existence of an "area" can be explained by the existence of the national people inscribed on it and, in turn, that this "people" can be explained by the area in which they are "discovered." This circular form of ascribing meaning is designed precisely to counter the flux of historicity itself, to elide and evade by reduction the combinatory and fluctuating partial determinacy of all social life. In doing so, area studies always avoids, and in any case fears, asking the questions of the history of the process by which specific difference was itself formed, in favor of presuming as its starting point the systematic logic of specific difference in its "civilizational" gradient as a fait accompli.

Historically, area studies was constituted in the shift from area as explanandum (something to be explained) to area as explanans (that which explains). This transition is important because it demarcates the fact that area is not here understood as a geographical determinism, but rather marks the boundaries of positionality of a social relation—this is why the logic of area is intimately related to the constitution of the West itself. That is, the West is not a geographical entity, ethnic category, or other substantialized entity. Rather, the West itself is fundamentally a social relation: the West is that position that observes the field or area, that catalogs, parcels, and apportions knowledge concerned with how a heterogeneous sequence of empirical singularities becomes "historically situated." What is at work in this operation is the continual transformation of territory into a field of power, a field onto which the "historically situated" sequence can be inscribed as a unitary people, for whom the field or area becomes that which explains them. This is the characteristic modality of area studies' coloniality of knowledge: a process of the naturalization and grounding of the flux of difference onto a transformed site, forming a new set of referral loops and feedback mechanisms enabling the circuit of explaining and explained to repeat such that the loop itself cannot be questioned from within it.

What is the aim of this capture? It flows from the primordial state desire, the operation that underpins as substratum all forms of the drive to simultaneously make-hierarchical and make-commensurable. It underpins the process of formation of the world as hierarchized commensurability and its continuing renewal: the state as a drive, as a desire for closure, appropriation, and categorization is an ideational flow; it is contained on the simultaneously vast and yet infinitesimal space of the border itself, constantly renewing its drive through the act of bordering. This flow of capture connects back onto itself, it feeds off its own origin by representing its own effects as if they were its origin, "it produces an image of identity as though this were the *end* of the different" (Deleuze 1968: 385; 1994: 301).

In this sense, area studies cannot be *overcome*, but must be bisected by a subtractive strategy of exodus, a removal of its basic element, which is ourselves. Such an exodus must be produced, not discovered, by creating new forms of encounter and theoretical practice, new lines of proximity, distance, and relation, the superimposition of outlines, boundaries, and operations in sites other than their expected location. This can only be done by folding area studies back onto itself through determined acts of recombination, in other words, by modifying "the relation between different elements in such a way as to produce a semiotic and practical effect different from that which determined the previous combination" (Berardi 2001: 199). Area studies is a form that is organized to prevent the disclosure of the operation of area as a flow of desire, as a desire for stable belonging. The work of recombination, the work of forcing the operation of area to disclose its function is a crucial aspect of formulating a biopolitical response to the world we inhabit, and the formulation of such a response can allow us to imagine new possibilities of figuration of ourselves, new possibilities of politics and sociality capable of entering fully into the historical present, when we are constantly being told that such alternatives are unavailable. What is at stake in the question of area studies, in the question of *area* itself, is nothing less than the historical potential to encounter ourselves without the comfort of the schema of civilizational difference, beyond the enclosure of the world that is implied in the violence of the form of belonging. This possibility can only emerge to the extent that we undertake the vast and endless labor of translation involved in excavating the operation of this uncanny historical force called "area."

This labor of translation, beyond the regime of predetermined and presupposed "difference" in which translation is simply represented as exchange, might also be called an ethics beyond enclosure, beyond the ironic modern "freedom" (the "free" individual as posited within specific difference) of being *included* by capital and national language. We cannot hope to undertake a strategy of exodus from the regime of civilizational difference that is this perverse form of "area" unless we grasp the betrayal of the expected arrangements, the expected forms of belonging, as an entirely affirmative act. In this sense, the labor of translation and critique against the logic of "area" is itself nothing less than the possibility of a transversal sociality beyond the quintessential modern form of capture, or *inclusion into the anthropological difference*: we must continue to encounter each other in this site of the production of knowledge, not as a negative multitude of stable subjects, but as an affirmative multitude of betrayers or traitors.

Notes

Early versions of parts of this essay were given as presentations at the Digital Archive and the Future of Transpacific Studies conference, held at Cornell University in Ithaca, NY, September 12–14, 2008, and the Rewriting Modern and Contemporary Japanese Intellectual History conference, held at Tôhoku University in Sendai, Japan, September 26–28, 2009, and I benefited from discussions with many of the participants. All translations from languages other than English are mine unless otherwise indicated.

1   For a crucial theoretical development of the articulation between the problem of "translation" and the question of "area," see the important text of Naoki Sakai and Jon Solomon, "Addressing the Multitude of Foreigners, Echoing Foucault" (2006).

2 Foucault's final period of work, in particular his investigations of the question of truth and the form of *parrhesia*, politicizes and extends his earlier analyses of the politicality of ethics, the ethics of the care of self, an ethics of the practical-social spacing between the "I" and the self that emerges in the topos of sociality, an ethics that is completely different from understandings of "sympathy" or "communication" across a putative divide. I attempt to discuss and cross-read this material later when I touch on the form of "area studies."

3 In an economic-historical sense, enclosure indicates on the one hand the process of the deterritorialization of the peasantry (and form of the rural village) from the land, but on the other hand also indicates something entirely different from this understanding of a direct violence, because what is deterritorialized is always also reterritorialized. That is, enclosure is in essence the process of formation, the creation, of the *owner* of this strange thing called "labor power," the commodity of labor power. In this sense, primitive accumulation is never a unilateral process of destruction or expropriation, but an act of creation. This point might be considered obvious, but its theoretical significance is often elided in certain contemporary analyses of the question of enclosure today.

4 This analysis is expanded in Gavin Walker, "The Schema of the West and the Apparatus of Capture: Variations on Deleuze and Guattari."

5 On the Marxian reading of Carl Schmitt's understanding of the term *katechon* (restraining, holding back), see Walker 2010.

6 See the lecture of March 1, 1978, in Foucault 2007. Let me note also that Uno and Foucault's concern with the question of the conduit, the question of "passing through" (*tôru*; *tôsu*; *conduire*), is deeply related to the problem of understanding the split between translation as a practice and its representation.

7 On the question of "semblance" (*Schein*; *kashô*), see the decisive point made by Uno (1973a).

8 Thanks to Ken Kawashima for alerting me to Gidwani's interesting work and for discussions related to this point.

9 I extensively analyze the theoretical relations between Schmitt's discussion of the *Ur-Akt* of the *nomos* and Marx's grasp of the process of primitive accumulation in Walker 2011.

10 We should note the discursive economy of this sequence of *setzen*, *besetzen*, *übersetzen*, and so on. That is, we should always recall the primordial element of "positing" (*Setzung*) that underpins every investment (*Besetzung*) and translation (*Übersetzung*), which always recurs back to its starting point, a paradoxical circuit of "presupposition" (*Voraussetzung*).

11 Here, I would also note that an essential analysis of these letters can be found in Bruno Bosteels 2011.

12 In general, on the politicality of this operation and its exigency today, see Sakai 2007, esp. 293–306.

13 Clearly Agamben's understanding of "courage" is deeply related to Foucault's final work on "the courage of truth." See Foucault 2009.

References

Agamben, Giorgio. 2002. *Remnants of Auschwitz: The Witness and the Archive*. Translated by Daniel Heller-Roazen. New York: Zone.

Agamben, Giorgio. 2008. *Che cos'è il contemporaneo?* Rome: Nottetempo.

Berardi, Franco (Bifo). 2001. *La fabbrica dell'infelicità: New Economy e movimento del cognitario*. Rome: Derive Approdi.

Berardi, Franco (Bifo). 2004. *Il sapiente, il mercante, il guerriero: Dal rifiuto del lavoro all'emergere del cognitario*. Rome: Derive Approdi.

Bosteels, Bruno. 2011. *La révolution de la honte: Autour de la correspondance de Karl Marx et Arnold Ruge*. Paris: La Fabrique.

Deleuze, Gilles. 1968. *Différence et répétition*. Paris: Presses Universitaires de France.

Deleuze, Gilles. 1994. *Difference and Repetition*, translated by Paul Patton. New York: Columbia University Press.

Deleuze, Gilles, and Félix Guattari. 1980. *Mille Plateaux*. Paris: Minuit.

Deleuze, Gilles, and Félix Guattari, 1987. *A Thousand Plateaus: Capitalism and Schizophrenia*, translated by Brian Massumi. Minneapolis: University of Minnesota Press.

Foucault, Michel. 1976. "Droit de mort et pouvoir sur la vie." In *La volonté de savoir*. Vol. 1 of *Histoire de la sexualité*, 177–211. Paris: Gallimard.

Foucault, Michel. 1996. "The Ethics of Pleasure." In *Foucault Live: Collected Interviews, 1961–1984*, edited by Sylvère Lotringer, 371–82. New York: Semiotext(e).

Foucault, Michel. 2007. *Security, Territory, Population*, translated by Graham Burchell. London: Palgrave. Originally published as *Sécurité, territoire, population: Cours au Collège de France, 1977–1978* (Paris: Seuil/Gallimard, 2004).

Foucault, Michel. 2009. *Le courage de la verité: Le gouvernement de soi et des autres II, cours au Collège de France, 1984*. Paris: Seuil/Gallimard.

Gidwani, Vinay. 2008. *Capital, Interrupted: Agrarian Development and the Politics of Work in India*. Minneapolis: University of Minnesota Press.

Guattari, Félix. 2009. "Capitalism and Schizophrenia." In *Chaosophy: Texts and Interviews 1972–1977*, 53–68. New York: Semiotext(e).

Marrati, Paola. 2001. "Against the Doxa: Politics of Immanence and Becoming-Minoritarian." In *Micropolitics of Media Culture*, edited by Patricia Pisters, 205–20. Amsterdam: Amsterdam University Press.

Marx, Karl. 1975a. "Difference between the Democritean and Epicurean Philosophy of Nature." In Vol. 1 of *Collected Works of Karl Marx and Frederick Engels*, 25–108. Moscow: Progress.

Marx, Karl. 1975b. "Letter of March 1843 to A. Ruge." In Vol. 3 of *Collected Works of Karl Marx and Frederick Engels*, 133–34. Moscow: Progress.

Marx, Karl. 1975c. "Economic and Philosophic Manuscripts of 1844." In Vol. 3 of *Collected Works of Karl Marx and Frederick Engels*, 229–348. Moscow: Progress.

Minatomichi, Takashi. 1998. "Kioku shikari, bōkyaku shikari" ("Hail Memory, Hail Forgetting"). *Gendai shisō* 26, no. 14: 84–113.

Nancy, Jean-Luc. 1983. *La communauté désœuvrée*. Paris: Christian Bourgois.

Nancy, Jean-Luc. 1991. *The Inoperative Community*, edited by Peter Connor. Minneapolis: University of Minnesota.

Nietzsche, Friedrich. 1980. "Unzeitgemässe Betrachtungen II: Vom Nutzen und Nachtheil der Historie für das Leben." In *Kritische Studienausgabe* (KSA). Vol. 1. Edited by Giorgio Colli and Mazzino Montinari, 243–334. Berlin: de Gruyter.

Sakai, Naoki. 2007. *Nihon/eizō/beikoku: Kyōkan no kyōdōtai to teikokuteki kokuminshugi* (*Japan/Image/America: The Community of Sympathy and Imperial Nationalism*). Tokyo: Seidosha.

Sakai, Naoki. 2008. *Kibō to kempō: Nihon kokukempō no hatsuwa shutai to ōtō* (*Hope and Constitution: The Subject of Enunciation of the Japanese Constitution and its Response*). Tokyo: Ibunsha.

Sakai, Naoki. 2009. "Translation and the Schematism of Bordering," translated by Gavin Walker and Naoki Sakai, paper presented at Gesellschaft übersetzen: Eine Kommentatorenkonferenz, Universität Konstanz, Germany, October 29–31.

Sakai, Naoki, and Jon Solomon. 2006. "Introduction: Addressing the Multitude of Foreigners, Echoing Foucault." In *Translation, Biopolitics, Colonial Difference*, edited by Naoki Sakai and Jon Solomon. Vol. 4 of *TRACES: A Multilingual Series of Cultural Theory and Translation*, 1–35. Hong Kong: Hong Kong University Press.

Schmitt, Carl. 1974. *Der Nomos der Erde im Völkerrecht des* Jus Publicum Europaeum. Berlin: Duncker & Humblot.

Schmitt, Carl. 2003. *The* Nomos *of the Earth in the International Law of the* Jus Publicum Europæum, translated by G. L. Ulmen. New York: Telos Press.

Spivak, Gayatri Chakravorty. 1999. *Critique of Postcolonial Reason*. Cambridge: Harvard University Press.

Spivak, Gayatri Chakravorty. 2003. *Death of a Discipline*. New York: Columbia University Press.

Tazaki Hideaki. 2007. "Haji, ikari, sonzai" ("Shame, Anger, Being"). In *Munō na mono tachi no kyōdōtai* (*The Community of the Incapable*), 113–25. Tokyo: Miraisha.

Uno, Kōzō. 1973a. "Rōdōryoku naru shōhin no tokushusei ni tsuite" ("On the Specificity of the Labor Power Commodity"). In *Uno Kōzō chosakushū*, 486–505. Vol. 3. Tokyo: Iwanami Shoten.

Uno, Kōzō. 1973b. "Rōdōryoku no kachi to kakaku" ("The Value and Price of Labor Power"). In *Uno Kōzō chosakushū*, 123–42. Vol. 4. Tokyo: Iwanami Shoten.

Uno, Kōzō. 1973c. "Benshōhōteki mujun ni tsuite" ("On Dialectical Contradiction"). In *Uno Kōzō chosakushū*, 407–39. Vol. 10. Tokyo: Iwanami Shoten.

Walker, Gavin. 2010. "Kenryoku toshite no shihon: Seijiteki kake to kyō no kishōsei" ("Capital as Power: The Political Wager and the Rarity of the Commons), *Muri to iu shikii to kyō no seisan*, part 2. *Jōkyō* 96, no. 3: 185–203.

Walker, Gavin. 2011. "Primitive Accumulation and the Formation of Difference: On Marx and Schmitt." *Rethinking Marxism* 23, no. 3: 384–404.

Walker, Gavin. 2018. "The Schema of the West and the Apparatus of Capture: Variations on Deleuze and Guattari." *Deleuze Studies* 12, no. 2: 210–35.

Politics and Translation: Reflections on Lyotard, Derrida, and Said

Étienne Balibar
Translated from the French by Gavin Walker

Introduction

Since the dawn of modernity, there have been two competing "models" to think and represent politics: war (and more generally conflict, or struggle, as in Machiavelli) and commerce (in the broad sense this term had in the classical age, for example in Kant and Montesquieu). These two terms are obviously not entirely independent, but rather compete or even interfere with each other. With the political, social, and cultural transformations linked to globalization, from the expansion of communication to the encounter between cultures in a postcolonial context, in which all the old and new nations find themselves implicated in one sense or another, this antithesis has not disappeared but rather has taken on new forms. This must increasingly lead us to a reflection on the possibilities and obstacles for translation, both as an everyday practice in which millions of people are involved, a

positions 27:1 DOI 10.1215/10679847-7251832
Copyright 2019 by Duke University Press

vital institution for the exercise of power, and as a rich and complex theoretical problem. I will attempt to explicate this set of issues with the assistance of certain works of three great contemporary philosophers, all recently deceased—Jean-François Lyotard's *The Differend* (1983), Edward Said's *Orientalism* (1978) and *Culture and Imperialism* (1993), and Jacques Derrida's *The Monolingualism of the Other; or, The Prosthesis of Origin* (1996).

* * *

Rather than a lecture as such, what I would like to present here is a fragment of a work in progress. It happens that recently, as the subject of my teaching, I began to explore a question that I will tentatively call "War and Translation: Two Models of Politics" in terms of its historical antecedents and its contemporary reemergence.[1]

There are numerous points of contact and correspondences between this experimental program and what is described in the thematic outline of the "Thursdays of the Social Sciences," under the title "What is Translated? Translation and Comparison." This is the case in particular for everything that concerns the idea of *an anthropological concept of translation* (and not merely linguistic or philological), and for the interrogation of the relationship between translation and violence (the argument that translation *can* "help defuse violence," but that this is neither inevitable, nor its only possible outcome, regardless of the circumstances and the "genres of translation" considered). Our assumption is that these common interests have to do with the social transformations, tenuously unified under the rubric of "globalization," that eliminate the distinctions between the local and global contexts of translation and inscribe both the local and global in the register of a "cosmopolitics."

From the philosophical point of view at least, it first occurred to me that the opposition of war and translation has taken over an earlier opposition, decisive in the philosophy of the classical age, between *war and commerce*, sometimes thought as an opposition and other times as a complementarity. To demonstrate how this relation passes from one to the other, I refer to two passages from Jean-François Lyotard's *The Differend*, first published in 1983.[2] The first is section 218:

A phrase, which links and which is to be linked, is always a *pagus*, a border zone where genres of discourse enter into conflict over the mode of linking. War and commerce. It's in the *pagus* that the *pax* and the pact are made and unmade. The *vicus*, the *home*, the *Heim* is a zone in which the differend between genres of discourse is suspended. An internal "peace" is bought at the price of perpetual differends on the outskirts. (The same arrangement goes for the ego, that of self-identification.) This internal peace is made through narratives that accredit the community of proper names as they accredit themselves. The *Volk* shuts itself up in the *Heim*, and it identifies itself through narratives attached to names, narratives that fail before the occurrence and before the differends born from the occurrence. Joyce, Schönberg, Cézanne: *pagani* waging war among the genres of discourse.[3] (Lyotard 1988: 151)

The question posed here, that of the relation between politics and the figure of *the stranger* (which is neither "friend" nor "enemy" as such, but which contains within itself both virtualities), is illustrated by means of a spatial *schema* (or rather a *quasi-schema*, because here we are in a speculative domain, not that of the knowledge [*connaisance*] of nature, but that of the uncertain realities of history)—that of the "border." It is a question of knowing not only *what separates* a border but also *what occurs there*, what *passes* there [*ce qui s'y passe*], "on" the borderline [*«sur» son tracé*], immediately marked with the sign of ambivalence (encounter and conflict, exchange and violence, commerce and violence). But an earlier passage, inserted into the discussion of Kant's philosophy, and the problems that pose the unification of its two "regions" (one devoted to the theoretical problem of knowledge [*connaisance*], the other devoted to the practical problem of morality), shows that this schema is itself susceptible to a flux or variation:

Each genre of discourse would be like an island; the faculty of judgment would be, at least in part, like an admiral or like a provisioner of ships who would launch expeditions from one island to the next, intended to present to one island what was found (or invented, in the archaic sense of the word) in the other, and which might serve the former as an "as-if intuition" [in other words, as a concrete representation of substitution, an analogy] with which to validate it. Whether war or commerce, this interventionist force has no object, and does not have its own island, but it requires a milieu—this would be the sea—the *Archipelagos* or primary sea as the Aegean was once called. (Lyotard 1983: 190; 1988: 130–31)

With the substitution of a maritime "border" for a terrestrial "border" comes a modification in the schema. What is suggested here is not only that the borderline is disputed and uncertain but that the border itself ultimately cannot be found or located: no fixed line can be inscribed in the marine element unless it is fixed virtually by means of a map. Similarly, we have a distinction between the two opposed political notions of war and commerce, which is destined to remain uncertain. Indeed, it is particularly in the marine element that one finds this hybrid figure between the trader [*le commerçant*] and the warrior, the *pirate*, the public enemy of all states except when the states themselves utilize this figure in a more or less secret [*occulte*] fashion to "illegally" combine war with commerce.

The opposition of war and commerce is always susceptible to being inverted, made into a complementarity, or even to expressing a fundamental identity. This is why its alternate formulation—war and peace—is also essential and problematic in the modern era. From Saint Augustine to Kant, political philosophy has reflected this ambivalence. Sometimes political philosophy sees peace, in the heart of an empire, or within the framework of an "international order," as the condition of possibility for commerce. At other times it sees in commerce a moral and material factor for the establishment of peace. Montesquieu's famous expression "sweet commerce" [*«le doux commerce»*] is often cited in this regard. In *The Spirit of the Laws*, it refers to a broad and differentiated notion (one could say that it is precisely "anthropological" avant la lettre) of commerce, which includes both commodity exchange and relations of "civility" or of "politeness," especially as they were established by the initiatives of women within the framework of a civilization based around a court society (see Elias 1983).[4]

But the peaceful essence of commerce remains eminently dubious. It is not only that there are hybrid forms (we see here piracy or "privateering" [*«course»*]). But these roles are exchanged in certain circumstances, perhaps inevitably: there are commercial wars, and above all there are *wars for com-*

merce, to impose its "freedom" (a short modern history from the origins of liberalism to the European imperialist expeditions intended to *force open the "closed" empires* such as Japan or China in the nineteenth century, to "open their doors" to commerce, that is to say, to *competition [concurrence]*). This ambiguity was immediately seized on by thinkers such as Kant (who, in turn, inspired Lyotard in the passages cited above), who referred to the combination of reciprocity and antagonism that constitutes the "motor" of historical progress as an "unsocial sociability" (*ungesellige Geselligkeit*). Marx identifies this ambiguity as well when he insists on the two sides comprising the extension of commodity circulation: on the one hand, the creation of a cosmopolitical "universality" through the institution of a *general equivalent* that measures all values, and through it comparing all human endeavors; on the other hand, the *violent dissolution* of traditional communities and their corresponding cultures, as their commodity exchanges with the exterior intensify and are gradually monetarized.[5]

Let us return now to Lyotard's *schema*. The notion of the "differend" [*«différend»*] around which the book as a whole is organized, forms the object of a series of successive elaborations, gradually deepening our understanding by passing from an experiment in relative thought to the "demonstration" of the absolute "wrong" [*tort*] suffered by the victims of the various phenomena of extermination and, more generally, of "universal suffering" (an expression of Marx that refers to the proletariat).[6] One such demonstration is as follows:

> The plaintiff lodges his or her complaint before the tribunal, the accused argues in such a way as to show the inanity of the accusation. Litigation takes place. I would like to call a differend the case where the plaintiff is divested of the means to argue and becomes for that reason a victim. If the addressor, the addressee, and the sense of testimony are neutralized, everything takes place as if there were no damages. A case of differend between two parties takes place when the "regulation" of the conflict that opposes them is done in the idiom of one of the parties while the wrong suffered by the other is not signified in that idiom. For example, contracts and agreements between economic partners do not prevent—on the contrary, they presuppose—that the laborer or his or her represen-

tative has had to and will have to speak of his or her work as though it were the temporary cession of a commodity, the "service," which he or she putatively owns. This "abstraction," as Marx calls it . . . is required by the idiom in which the litigation is regulated ("bourgeois" social and economic law). In failing to have recourse to this idiom, the laborer would not exist within its field of reference, he or she would be a slave. In using it, he or she becomes a plaintiff. Does he or she also cease for that matter to be a victim? (1983: 24–25; 1988: 9–10)

This demonstration is followed by another soon after:

Would you say that interlocutors are victims of the science and politics of a language understood as communication to the same extent that the worker is transformed into a victim through the assimilation of his or her labor power into a commodity? . . . This is where the parallels end: in the case of language, recourse is made to another family of phrases; but in the case of work, recourse is not made to another family of work, recourse is still made to another family of phrases. . . . To give the differend its due is to institute new addressees, new addressors, new significations, and new referents in order for the wrong to find an expression and for the plaintiff to cease to be a victim. This requires new rules for the formation and linking of phrases. No one doubts that language is capable of admitting these new phrase families or new genres of discourse. The differend is the unstable state and instant of language wherein something which must be able to be put into phrases cannot yet be. This state includes silence, which is a negative phrase, but it also calls upon phrases which are in principle possible. A lot of searching must be done to find new rules for forming and linking phrases that are able to express the differend disclosed by the feeling [*sentiment*], unless one wants this differend to be smothered right away in a litigation and for the alarm sounded by the feeling to have been useless. What is at stake in a literature, in a philosophy, in a politics perhaps, is to bear witness to differends by finding idioms for them. (1983: 29–30; 1988: 12–13)

These formulations turn around the problem of translation and the "untranslatable" in many ways. They suggest a much broader use of the notion of translation, and consequently of language (or *idiom*), than that which is typically (officially) received. The "borders" in question here—with the effects of strangeness or *foreignness* [*d'étrangeté ou d'étrangèreté*] that are induced by their contours [*tracé*]—are not purely "national," they could be *social*, or *moral*, or *religious*.[7] The differend itself, in other words, *the absence of the possibility to discuss*, and first of all "to respond" or "to present one's case" to the other, comes first. Differends are induced within one language— in the official sense of the term—just as they are among *different languages* that are between them untranslatable. But Lyotard complicates this analogy by inverting it. He argues that it is possible *in language*, by its transformation through the invention of new idioms that make "translatable" what was not, to express all conflicts based on an absolute or radical "wrong" [*tort*], a "wrong" that somehow excludes those who suffer from humanity or community. Obviously, to say that differends can be expressed does not mean that they are thereby "resolved." Rather, as demonstrated at the end of the statement quoted above, it means that these differends *can become the object of a politics*, and arise as "subjects" of this politics, with the help of literature and philosophy.

What is important here is that Lyotard defines the differend as a conflict that arises not between "people" [*«gens»*] but between "phrases" themselves. The fundamental experience of these "phrases" lies in an "impossible" situation in which "one must link" [*«il faut enchaîner»*] things together (phrases, discourses, narratives, arguments) that lack a means of being linked [*enchaînement*]. What expresses this is precisely the fact that while there is an urgent need to establish or reestablish social communication, there is a simultaneous and fundamental *impossibility* of translation.

Of course, this is somewhat abstract. The genre of philosophy requires it. But the genre also allows us to go further by conjoining determinate historical situations with singular subjective experiences. I want to now take up two contemporary authors for whom the question of "translation," and generally the difference of languages [*idiomes*], was a fundamental concern: Jacques Derrida and Edward Said. This confrontation will be based on two texts of an autobiographical character that have a less than apparent difference on the question of the relationship of the individual to his or her "mother tongue" [*«langue maternelle»*], the one in which the fundamental

processes of education and culture take root, which is also in the most literal
sense the "language of the mother." The point of this confrontation follows
not only from the fact that Derrida and Said represent two divergent tendencies in contemporary "critique" but also from the fact that they undertake
engaged reflections on fundamental aspects of the "Mediterranean differend" that was engendered by colonization and that was prolonged beyond
its official end. We can see here an illustration of what Rada Ivekovic, in an
article in *Transeuropéennes*, has called "permanent translation" [*«traduction
permanente»*], an infinite process of subjectivation linked to the "dispossession of the self" that entails the "existential paradox" on which depends the
very possibility of communication (see Iveković 2002; 2007).

The text of Derrida in question here is his *Monolingualism of the Other*:

My attachment to the French language takes forms that I sometimes consider "neurotic." I feel lost outside the French language. The other languages which, more or less clumsily, I read, decode, or sometimes speak,
are languages I shall never inhabit. Where "inhabiting" begins to mean
something to me. And dwelling [*demeurer*]. Not only am I lost, fallen,
and condemned outside the French language, I have the feeling of honor-
ing or serving all idioms, in a word, of writing the "most" and the "best"
when I sharpen the resistance of my French, the secret "purity" of my
French, the one I was speaking about earlier on, hence its resistance, its
relentless resistance to translation; translation into all languages, including another such French. Not that I am cultivating the untranslatable.
Nothing is untranslatable, however little time is given to the expenditure
or expansion of a competent discourse that measures itself against the
power of the original. But the "untranslatable" remains . . . the poetic
economy of the idiom, the one that is important to me, for I would die
even more quickly without it, and which is important to me, myself to
myself, where a given formal "quantity" always fails to restore the singular event of the original, that is, to let it be forgotten once recorded. . . .
Word for word, if you like, syllable by syllable. From the moment this
economic equivalence—strictly impossible, by the way—is renounced,
everything can be translated, but in a loose translation, in the loose sense
of the word "translation." In a sense, nothing is untranslatable; *but in*

another sense, everything is untranslatable; translation is another name for the impossible. . . . How can one say and how can one know, with a certainty that is at one with oneself, that one shall never inhabit the language of the other, the other language, when it is the only language that one speaks, and speaks in monolingual obstinacy, in a jealously and severely idiomatic way, without, however, being ever at home in it? . . . But above all, and here is the most fatal question: how is it possible that this language, the only language that this monolingual speaks, and is destined to speak, forever and ever, is not his? How can one believe that it remains always mute for the one who inhabits it, and whom it inhabits most intimately, that it remains *distant, heterogeneous, uninhabitable, deserted*? Deserted like a desert in which one must grow, make things grow, build, and project up to the idea of a route, and the trace of a return, *yet another language*? . . . But I will be told, not without reason, that *it is always that way a priori*—and for everyone else. The language called maternal is never purely natural, nor proper, nor inhabitable. *To inhabit*: this is a value that is quite *disconcerting* and equivocal; one never inhabits what one is in the habit of calling inhabiting. There is no possible habitat without the difference of this exile and this nostalgia. Most certainly. That is all too well known. But it does not follow that all exiles are equivalent. From this shore, yes, *from this* shore or this common drift, all expatriations remain singular. For there is a twist to this truth. This *a priori* universal truth of an essential alienation in language—which is always of the other—and, by the same token, in all culture. This necessity is here re-marked, therefore marked, and revealed one more time, still one more first time, in an incomparable setting.[8] (Derrida 1996: 97–114; 2006: 56–58)

A full interpretation of these lines would doubtlessly situate them in relation to the entire linguistic and philosophical tradition at the origin of which appears the name of Humboldt and at the destination, that of Benveniste. This tradition poses the question of *subjectivity in language* in terms of reciprocal appropriation: the appropriation of the subject by language, at least the one which is for him or her "maternal" or "first" and the appropriation of language by the subject, through the first impregnation relayed through learning [*l'apprentissage*], which is also part of translation. This tradition inscribed these appropriations within a "community" of interlocutors, who each trace the "borders" of membership (of the nation, but also of class, or membership in a culture), making the *mutual comprehension* of a "shared language" between these interlocutors the means of passing or repassing endlessly from I to We, from We to I (see Benveniste 1966–1974). What, then, does Derrida mean when he writes (and repeats several times—it is the enigmatic driving thread of his essay), "I only speak one language, and yet this (unique) language is not mine" if "the language of the other" does not mean "a foreign language" («*une langue étrangère*») but rather concerns the foreignness [*étrangèreté*] and strangeness [*étrangeté*] of the *mother tongue?*

In the text, the response to this question combines three aspects, each of which is situated at a different level of experience. The first aspect is *biographical*, marked by a dramatic episode in the childhood of the author, a French Jew in Algeria: the loss of the title of citizen for the Jews of Algeria, which they had received with the Crémieux Decree of 1870 to distinguish them from the other "natives" [«*indigènes*»], by the collaborationist Vichy government, taking advantage of the defeat to institute anti-Semitic legislation (see Stora 2006).[9] This loss resulted in exclusion from the public schools, which in France are principally concerned with the training of linguistic correctness and the internalization of language as the principle institution of national identity.

Here we encounter the second aspect, which is *cultural*. The French language, with which Derrida tells us he shares a "neurotic" or ultra-perfectionist "purist" relation, is an imperial (colonial) language, protecting itself from contamination by other "inferior" or "subaltern" idioms, whether from linguistic minorities in the metropole, or colonized populations, who are, contradictorily, compelled both *to assimilate* and *to stay in their place*, to remain dominated, and thus held in a situation of *internal exclusion*. It therefore rests implicitly on a "communitarian" hierarchy that is simultaneously the object of a powerful denial, the counterpart of which is the "haunting" of the dominant language by dominated languages and the more or less vast permeability of its expressions of social class.

To conclude, here we see also a *philosophical* aspect that we can call "transcendental" in the technical sense, since it concerns the *conditions of possibil-*

ity for the subject's access to language, which conditions its insertion into the world. Instead of instituting an equality, which would put into a simple relation the speaking individual with a community in which he or she naturally "inhabits" language, it forms a relation to language as much "expropriated" as it is "appropriated" (here Derrida coins the portmanteau term *exappropriation* to express this contradiction), that refers itself back to difficulties and to permanent conflict. The subject is "at war" with language, and fundamentally, through the subjects who utilize it, it is language that is "at war" with itself, with its own [*propre*] instituted existence (Derrida 1996: 63–64). But the representation of this internal conflict also opens onto a constructive practice: that of translation. A subject or speaker (*locuteur*) who is not in a natural relation of belonging with his or her "mother" tongue is always already inscribed into a process of "translation" of his or her own (*propre*) language, a process made all the more difficult precisely because there are no rules or codes for it. This is what Derrida calls "absolute translation" (61), which operates in relation to all others—that is to say in relation to all other "encounters" with foreign languages—both as a predisposition and as an obstacle.

At this point it becomes interesting to outline a confrontation with Edward Said, who draws from personal experience in the opposite direction to give us a proposition that converges in part with the relations of politics to translation. In his autobiography, Said—son of a Palestinian father who acquired US citizenship but returned to settle in Egypt and a Lebanese mother from a family of Baptist ministers, with an English first name and an Arabic last name—recounts the impossibility in which he found himself as he tried to determine which of the two languages as a child (English or Arabic) was "his language," even though, as shown in the following passage, the two languages had never been interchangeable:

> The travails of learning such a name were compounded by an equally unsettling quandary when it came to language. I have never known what language I spoke first, Arabic or English, or which one was really mine beyond any doubt. What I do know, however, is that the two have always been together in my life, one resonating in the other, sometimes ironically, sometimes nostalgically, most often each correcting, and commenting on,

the other. Each can seem like my absolutely first language, but neither is. I trace this primal instability back to my mother, whom I remember speaking to me in both English and Arabic, although she always wrote to me in English. . . . Certain spoken phrases of hers . . . were Arabic, and I was never conscious of having to translate them or . . . knowing exactly what they meant. They were a part of her infinitely maternal atmosphere . . . promising something in the end never given. But woven into her Arabic speech were English words like "naughty boy" and of course my name, pronounced "Edwaad." I am still haunted by the memory of the sound. . . . Her English deployed a rhetoric of statement and norms that has never left me. Once my mother left Arabic and spoke English there was a more objective and serious tone that mostly banished the forgiving and musical intimacy of *her* first language, Arabic. . . . I hadn't then any idea where my mother's English came from or who, in the national sense of the phrase, she was: this strange state of ignorance continued until relatively late in my life. (Said 2000: 4–5)

Here we have another form of uncertainty, one that is more "objective" in a sense, but which also rejects the necessity of translation as the condition for access to language in an unconscious zone that precedes the formation of individual consciousness. In his critical and political works, Said gradually worked out the conflictual and cultural dimensions of this uncertainty, which takes the form of an *unstable* relation between the dominant and dominated languages of the "imperial" world, specifically, that of the domination and incorporation of "Oriental" cultures, in particular those of the Southern Mediterranean, within the space historically and politically dominated by Euro-American culture. This evolution in Said's thought is particularly visible when one compares his formulations related to the *representation of the non-European other* in his 1978 *Orientalism*, which brought him fame but also embroiled him in a far-reaching polemic, with the essays of 1981 to 1987 on the treatment of Islam by the American media (Said 1997), and the essays in historical and literary criticism collected in *Culture and Imperialism* ([1993] 1994). Between these two moments the Iranian Revolution and the First Gulf War took place, as did the Oslo Accords on the "settlement" of the Palestinian question, to which Said, a member of the Palestinian National Council, was opposed from the outset, predicting that the accords would essentially be used by the State of Israel to intensify its colonization, and to instrumentalize the Palestinian Authority, whose formation Israel had foreseen to manage parts of the occupied territories after the 1967 War (see Said 1999).

In *Orientalism*, the theory of "hegemony" and its reproduction beyond the moment of formal decolonization rests on the detection of a *double asymmetry: between those who "represent" and those who are "represented," and between those who "translate" and those who are "translated."* Along with idioms of different status come the *narratives* (historical, literary) that assign a more or less mythical personality to peoples and cultures, as well as the *sciences* (or the academic disciplines: history, archaeology, philology) that allow them to be managed. But in his later works things become complicated because Said is confronted with the fact that the discourse of domination can be *reappropriated* by the "dominated" themselves, who use it either to express demands for autonomy, or to overturn and expose imperialism to its image of itself. Something else is added to the "cognitive" dimension of discourse, understood in the terminology borrowed from Foucault as what Said calls an apparatus [*dispositif*] of "power-knowledge," something more "dialogic" but always conflictual, in which the representations of East and West produce effects far more *ambivalent* than can be seen in the simple schema of colonial domination. *Nationalism* and especially the *politico-religious discourse* (Islamic "fundamentalism") are always susceptible to co-optation, since their own representation of the "Orient" as a world of irrational values fundamentally incompatible with those of the "Occident" regenerates an essential "untranslatability" between the supposed two halves of the world.

Said's critique, which is still controversial today, operates simultaneously *on two fronts*, which are paradoxically bound together in a common task of demystifying the stereotypes of cultural difference. This critique found a privileged expression in his reading of the work of Frantz Fanon, the psychiatrist from Martinique who joined the cause of Algerian independence, and died in 1961, developed particularly in *Culture and Imperialism*. This reading notably focuses on the *interiorization of cultural conflict* by each of the parties that occupy the space of colonization (and beyond it, into the "postcolonial" period), and on the simultaneously assymetrical and reciprocal possibilities for transformation contained in this structure:

One has the impression in reading the final pages of *The Wretched of the Earth* that having committed himself to combat both imperialism and orthodox nationalism by a *counter-narrative* of great deconstructive power, Fanon could not make the complexity and anti-identitarian force of that counter-narrative explicit. But in the obscurity and difficulty of Fanon's prose, there are enough poetic and visionary suggestions to make the case for *liberation as a process* and not as a goal contained automatically by the newly independent nations. Throughout *The Wretched of the Earth* (written in French), Fanon wants somehow to bind the European as well as the native together in a new non-adversarial community of awareness and anti-imperialism. (Said [1993] 1994: 274)

The unity of the "representation of the other" and of "assymetrical (or *unequal*) dialogue" is precisely what Said also calls *interpretation*. Its subjects or agents are not individuals so much as they are historically and institutionally constituted "communities of interpretation" located within the framework of a certain distribution of power between the various parts of the world and of humanity. But this distribution gives way to a play of power, a dislocation of the relations of forces: in the face of the hegemonic disciplines and established discourses, which reproduce the stereotypes of alterity, we discover the existence of "antithetic interpreters" [*«dissidents de l'interpretation»*] who produce effects of contestation and nurture resistance by modifying the regime of translation (Said 1997: 135). And in other essays, Said came to emphasize the analysis of the *condition of exile* in its multiple modalities as a situation that, while not in itself sufficient to *produce* a dissidence of interpretation, nevertheless has facilitated the boldest accomplishments throughout history, by destabilizing the adherence of subjects to a unique community blessed with "evidence" and "authority," and by installing itself in them up to an unstable and uncomfortable point where translation is simultaneously necessary, everyday, and disconcerting. We thus discover here a lesson comparable to that proposed by Lyotard in terms of the "differend" and by Derrida in terms of "exappropriation," despite all the divergences that separate these authors from each other. It is precisely on this "lesson" that I propose to continue working in the future.

Notes

[Trans.] The present essay was originally given as a presentation, "Politique et traduction: Reflexions à partir de Lyotard, Derrida, Said" at the seminar series "Que veut dire traduire? Traduction et comparaison—Les jeudis des sciences sociales" at the Université Jean Monnet in Saint-Étienne, on November 19, 2009. The translator would like to thank Stephen Hastings-King, as well as Étienne Balibar, for their corrections and suggestions.

1   Notably in October 2009 at the University of Chicago's Center for Contemporary Theory, which did me the honor of inviting me to give a series of eight seminars intended for doctoral students in anthropology and political science.

2   To understand the sense in which Lyotard takes up this term (he did not invent it, but expanded its use), the American translators of the English edition introduce a subtitle to this work: "Phrases in Dispute" (Lyotard 1988). A dispute is more than a litigation or a divergence, but less than a conflict or battle, and notably it does not necessarily take the binary form opposing two enemies or adversaries.

3   The etymologies suggested by Lyotard might be disputed: *paganus*, translated later by *paysan* (peasant) and by *païen* (pagan), comes from *pagus*, but this latter term does not mean *confines* so much as *village* in classical Latin and does not have, as it appears, an etymological relation with *pax* (peace) or with *pactum* (pact, treaty). But one can take it in its spirit as a simple wordplay.

4   On "sweet commerce" in Montesquieu, see Spector 2004. The complexity of the semantic specter of "commerce" in the classical age can be located in the English *intercourse* and in the German *Verkehr*.

5   Dipesh Chakrabarty, in his *Provincializing Europe* (2000), adopts this schema to describe two models of "translation" and thus two modalities of relation to language and to the diversity of languages.

6   It is also linked to a sharp debate at the time around the "revisionism" concerning the "absence of material proof" for the existence of gas chambers in the Nazi extermination camps (against which the "survivors" are somewhat impoverished, precisely because they have only their "subjective" testimony [*témoignage*] to pose against it).

7   Note that I do not say "cultural," precisely because the examination of the question of the differend is one of the means we have to *put into question* the notion of "culture" itself, hitherto promoted by anthropology, which is in the process of being renounced here.

8   This "incomparable setting" mentioned at the end of the excerpt is that of Louisiana (the University of Baton Rouge), where the conference for which Derrida prepared the first version of this text was held.

9   We must recall, as Derrida insists, that the German troops did not occupy Algeria—the government of the French state was not obeying in this case any external coercion, but following their own tendencies.

References

Benveniste, Émile. 1966–1974. "La subjectivité dans la langue." In *Problèmes de linguistique générale*. 3 vols. Paris: Gallimard.

Chakrabarty, Dipesh. 2000. *Provincializing Europe*. Princeton, NJ: Princeton University Press.

Derrida, Jacques. 1996. *Le monolinguisme de l'autre ou la prothèse d'origine*. Paris: Galilée.

Derrida, Jacques. 2006. *The Monolingualism of the Other; or, The Prosthesis of Origin*, translated by Patrick Mensah. Stanford, CA: Stanford University Press.

Elias, Norbert. 1983. *The Court Society*. New York: Pantheon.

Fanon, Frantz. 1961. *Les damnés de la terre*. Paris: F. Maspéro.

Fanon, Frantz. [1963] 1965. *The Wretched of the Earth*, translated by Constance Farrington. New York: Grove Press.

Ivekovic, Rada. 2002. "De la traduction permanente (Nous sommes en traduction)." In "Traduire, entre les cultures." Special issue, *Transeuropéennes: Revue internationale de pensée critique*, no. 22: 121–45.

Ivekovic, Rada. 2007. "Langue coloniale, langue globale, langue locale." *Corpus: Revue du Collège Internationale de Philosophie*, no. 58: 26–36.

Lyotard, Jean-François. 1983. *Le différend*. Paris: Minuit.

Lyotard, Jean-François. 1988. *The Differend: Phrases in Dispute*, translated by Georges van den Abbeele. Minneapolis: University of Minnesota Press.

Montesquieu, Charles de Secondat. [1748] 1969. *De l'esprit des lois*. Paris: Éditions sociales.

Montesquieu, Charles de Secondat. 1989. *The Spirit of the Laws*. Edited by Anne M. Cohler, Basia C. Miller, and Harold S. Stone. Cambridge: Cambridge University Press.

Said, Edward. 1978. *Orientalism*. New York: Vintage Books.

Said, Edward. [1993] 1994. *Culture and Imperialism*. New York: Vintage.

Said, Edward. 1997. *Covering Islam: How the Media and the Experts Determine How We See the Rest of the World*. New York: Vintage.

Said, Edward. 1999. *Palestine: L'égalité ou rien*. Paris: La Fabrique.

Said, Edward. 2000. *Out of Place: A Memoir*. New York: Vintage.

Spector, Céline. 2004. *Montesquieu: Pouvoirs, richesses, et sociétés*. Paris: Presses Universitaires de France.

Stora, Benjamin. 2006. *Les trois exils, juifs d'Algérie*. Paris: Stock.

The Hidden Area between Marx and Foucault

Ken C. Kawashima

> Explaining things from below also means explaining them in terms of what is most confused, most obscure, most disorderly, and most subject to chance. . . . [It] is the obscurity of contingencies and all the minor incidents that bring about defeats and ensure victories.
> —Michel Foucault, *"Society Must Be Defended"*

Introduction

When, in the lectures at the Collège de France, Foucault (1981: 151) speaks of the essential phenomena that biopolitical power addresses, it is never simply "populations" but rather those phenomena that are most aleatory, contingent, and subject to chance within these populations. He writes, "The phenomena addressed by biopolitics are, essentially, aleatory events that occur within a population that exists over a period of time." Biopolitical power does not simply represent, discipline, and police populations. It represents, disciplines, and polices that which is most resistant within a population to a logic of necessity that is most prone to chance, the hazardous accident, and contingency. Moreover, it does so through the use of force on the somatic body as well as on the mind. While this discourse on the aleatory is commonly associated with a long and diverse line of thinkers—from Cage, Althusser, Deleuze,

positions 27:1   DOI 10.1215/10679847-7251845
Copyright 2019 by Duke University Press

and Badiou to Mallarmé, Kuki Shuzo, Nietzsche, Hegel, and Epicurus—
the discourse of the aleatory event is also profoundly embedded in Marx's
critique of political economy at the heart of his analysis on the abstract laws
of motion of the "capitalist mode of production," which, alongside its attendant "forms of intercourse" or exchange, repress historical contingencies of
the past within its inexorable logic of necessity in the present. Let us begin,
therefore, by discussing three areas in Marx's key text of *Capital* in which
the problem of the aleatory is clearly foregrounded. We will then turn to
Foucault's distinct approach to the matter.

For Marx, the analysis of the aleatory first emerges—as an initial posing
of his own version of the dialectic of necessity and contingency—in the first
chapter of *Capital*, where he analyzes the commodity form, its unrepresentable contradiction between use value and exchange value, and the unfolding
of the forms of value from commodity to money to capital. From the perspective of the seller, the aleatory event is found in the leap from position of
the commodity (C) to the position of money (M), or C-M. Karatani Kojin,
also deeply fluent in the discourse of the aleatory, was arguably the first to
remind us that the aleatory event in Marx is found in the experience of contingency and the risk that is inherent to the act of selling. This is what Marx
calls the salto mortale.[1] As Karatani (1985: 116) has argued, the position of
selling is structurally always weaker than the position of buying precisely
because the salto mortale only exists on the side of the seller, who must pray,
so to speak, to realize value in a sale, without which he would be incapable
of purchasing daily necessities. From the perspective of the seller, an unrepresentable chasm separates the seller from money. While money logically
derives from the circulation of commodities, the form of money (and what
Marx calls "money as money") ascends to a meta level of representation,
appearing now as a universally sublime and monotheistic object of fetishistic
transvaluation. Compared to the position of selling, the position of buying
has more power because it is structurally in a position determined by the
money form—the "universal equivalent"—that can be exchanged with any
other particular commodity sans hazard.

A second area on the aleatory in *Capital* was pointed out by the late
Althusser in his essay on aleatory materialism.[2] Here, Marx discusses the
aleatory in relation to the origins of the capitalist mode of production and

the manner in which the violence of these origins are represented and repressed by the narrative of so-called "primitive accumulation": the liberal narrative that disavows the actual histories of the deployment of force by the state that dispossessed the masses, a disavowal at the heart of the creation of a national labor market and bourgeois ideology.[3]

It also is important to recognize that these two instances of the aleatory event in *Capital* operate on distinctly different levels of analysis, revealing an obscure area in Marx's method that becomes especially visible in the important sections on the accumulation process. The first level then is centered in Marx's demonstration of capital's inner logic, which unfolds out of the abstract category of the commodity form, as if by a miracle of the work of the negative. By contrast, the second level is related to the historicity and subsequent historicization and periodization of capitalism's origins and development that revolve around state policy, which for both Marx and Foucault would include economic and social state policies. With this second level of analysis, Marx thus critically follows the state's discourses on property, poverty, homelessness, vagrancy, and so on, all of which can be summed up by Marx's phrase "bloody legislation." On this level Marx produces a counterdiscourse of these origins in order to smash the moralizing fantasy and discourse of frugality, abstinence, and self-discipline.

Mediating these two instances of the aleatory in Marx's narrative is a third aleatory event, namely capitalist crisis, which always appears suddenly yet inevitably and which radically splits the sphere of circulation from the sphere of production, leaving as its residue what Marx (1991: 359) calls "excess capital" co-existing alongside surplus populations, the phenomenal result, as well as the precondition, of cyclical capitalist crisis. Visible especially in phases of recession, this contradiction or splitting has all the characteristics of something that requires nothing short of—again—a miracle to suture it back together. In crisis, all that is in motion in capitalist circuits comes to a halt, arresting both spheres of production and circulation. In the depths of crisis and recession, sellers and buyers alike—perhaps for once—experience an uncanny equality, for both are minimally miserable for the same cause: a profound lack of money, the universal equivalent. In phases of crisis and recession, the distinction between buyers and sellers is suspended—ephemerally or indefinitely—which cannot avoid arresting the work of the norms of the working day. In crisis and recession, as Marx has shown, it is precisely this norm of the working day that is reorganized and recalibrated, with different and new ratios of constant and variable capital in the "organic composition of capital," for a new cycle of production (which is already on its way to crisis). In the depths of crisis and recession, however, the norm is not stable but is thrown into question while masses wait anxiously in the margins of productive capital and miserably in the sphere of circulation, where, as Marx reminds us, "Between equal rights, force decides" (*Capital*, vol. 1: 344).

What especially requires further specification in Marxist scholarship is how the question of force can be further elaborated in relation to the three aleatory events of capitalism, events that lead to either the smooth reproduction of relations of production or to its permanent suspension and eradication. It is in this context that Foucault's lectures on biopolitics are valuable and helpful for dialectical and historical materialist analysis, which should be supplemented by Foucault's logic of strategy (Foucault 1981: 44). In complex and powerful ways, Foucault's lectures on biopolitics can be read as a sustained effort to specify the nature of these aleatory events as they relate to the historical emergence and reproduction of what Marx called in the preface to *Capital* the "capitalist mode of production and the forms of intercourse corresponding to it."

Foucault's "stand" on Marxism, which was at times clearly "standoffish," is one in which he prefers to isolate what he calls "certain real problems" that fall outside Marx's analysis of the motion or the logic of capital but that also exist within it, albeit in a hidden way. As Foucault (151, 156) said in an interview with Duccio Trombadori, "If one wants the great systems finally to be open to certain real problems, it is necessary to look for the data and the question in which they are hidden."

In terms of Marxist discourse, we can say that the problems Foucault analyzes are hidden between the two levels of analysis that Marx's narration in *Capital* reveals: (1) the ontological (or onto-theological) level of capital—the commodity form—and the totality of the interrelated laws of motion of the capitalist mode of production, and (2) the level of state discourse and ideology ("bloody legislation"). Put differently, Foucault's genealogies dwell between the processes so clearly identified by Marx of *exploitation* (explained on the axiomatic first level of logic) and *expropriation* (explained on the second level of the state). Biopolitical (and disciplinary) power exists in the interstices of expropriation and exploitation, mediating and distorting this relation with their own techniques of power, which are carried out on a microphysical, institutional scale.

Seen from this perspective, Foucault's analysis of biopolitics is already structured by Marx's critique of political economy. When Foucault (2003: 14) asks in his lectures from 1975 whether power is "modeled on the commodity," and when he implicitly says "no" to this question, the question we need to ask today is, if power is not reducible to the commodity form, then how, at least, does power/knowledge produce the conditions for—or against—the emergence of the commodity form and the dialectical logic of capitalist (re)production and accumulation? How can power/knowledge, instead of facilitating and bridging the relation between capitalist expropriation and exploitation, do the opposite work of permanently disabling and eradicating it?

In our contemporary conjuncture of post–Cold War, post-9/11 ideologies of capitalist triumphalism, the epistemic uses and abuses of reading Foucault as a theoretico-political alibi to ignore, dismiss, or disavow Marx's analysis of the logic of capital have to be identified plainly for the intellectual, political, and even cultural bankruptcy that they too often have become. Thus, instead of arguing how Foucault's analysis of biopower can, in fact, be isolated from the Marxist analysis of the commodity, we should instead locate where Foucault's analysis of biopolitics necessarily leads us to focus on the relations between (1) power/knowledge as a question of strategy, (2) the conditions of emergence of the commodity form, and (3) the dialectical logic of capitalist accumulation and reproduction that develops out of this form. If Foucault's analysis of biopolitical power is not reducible to the commodity—if it is outside the problem of the commodity as he says—then it is clear that this order of biopolitical power will at least have a relation of separation to the commodity form. It is this relation of separation, which is also an actual material relation of conjunction or disjunction, that needs to be considered après Foucault precisely so that we know where to draw the line when it comes to the disorienting and obscure chaos of "force against force."

Put differently it is arguable that, in this *milieu* or *area* between power/knowledge and the commodity form, the capitalist mode of production is most vulnerable to an explosive, conjectural condensation and concentration. At the same time, as a milieu, it is something that is constantly in movement, frustrating efforts to define and represent it let alone to organize an actual concentration of forces out of it. Yet, for this very reason, what kind of political orientation and direction must be given to this milieu beyond the paltry and apologetic representations of it that predominate not only discursively but also ideologically?

Here we must introduce the importance of a shared discursive term between the methods of Foucault and Marx: the *floating population*. Both Foucault and Marx show, in their differing ways, that when it comes to representing the specific category of the floating populations, there is always a distortion of the truth that is fueled by the interests of capital, the state, and the apparatuses of power, which rush to repress the living voices of the floating populations through administrative, bureaucratic, statistical, and social scientific expertise, archiving, and policing. Both Marx and Foucault want to think beyond existing (especially liberal) modes of representing and policing the present situation, smashing them (and their methods) in order to finally establish and institute entirely new modes of thinking and being, entirely new modes of mass organization that soberly yet irrepressibly address and push through the bitter and real contradictions of society, which continue to be repressed and disavowed precisely through these modes of representing populations, who never asked to be represented in the first place.

Foucault's analysis of how biopolitical power targets the "aleatory events occurring within a population" helps us isolate and understand that which represents the most vulnerable process for the world of capital. This aleatory process is not found in populations, per se, but rather in relation to the peculiar commodity of labor power that human populations, subsumed beneath the despotism of capital, are compelled to embody, possess, bear, and sell, for it is the most indispensable commodity for capitalist production. The contradictory character of labor power as a commodity, however, is that it cannot be produced as a commodity directly. It must thus be captured indirectly from outside its world, a detour that deviates from capital's ideal path, which presupposes the availability of labor power. The process of the com-modification of labor power is thus a fundamental process of capitalism that overcomes the "original" yet repeated fact that capital cannot produce labor power as a commodity directly, yet is constantly compelled to consume it in order to produce new values. *Capital*'s Achilles' heel is exposed in this con-tradiction surrounding labor power, and yet we also can say it is repressed by an order of power/knowledge that Foucault illuminates through genea-logical analysis.

The Floating Signifier of Populations

Where Foucault and Marx encounter the real of this repressed truth of the capitalist order, but where they also cleave divergently from each other, is around the figure and concept of the floating population, which discursively floats between them like an unmoored signifier. This is true especially for Foucault, who is fundamentally critical of ever speaking for, or on behalf of, floating populations. Instead, he follows the surface representations of these populations as a problem in relation to, it is said in biopolitical discourse, "society must be defended." Born within concrete institutional milieux, these discourses reveal vast, interconnected networks of power/knowledge, a genealogy of power irreducible to sovereign state power. As Foucault writes, "Example of the army: We may say that the disciplinarization of the army is due to its control by the state. However, when disciplinarization is con-nected . . . with the problem of floating populations, the importance of com-mercial networks, technical inventions . . . a whole network of alliance, sup-port, and communication constitutes the 'genealogy' of military discipline (Foucault 2007: 162)."

Who is calling floating populations a problem, and for whom is the prob-lem of floating populations meaningful and full of value? For Foucault, it is meaningful within a larger, more expanded circulation of signs of the economy, which delineate the contours and norms of those most directly invested in the movements of the market. For example, in *The Birth of Bio-politics*, Foucault discusses post-Keynesian discourses of "absolute poverty" and speaks of a floating population that "can always be available for work, if market conditions require it." He writes:

A kind of infra- and supra-liminal floating population, a liminal population which, for an economy that has abandoned the objective of full employment, will be a constant reserve of manpower which can be drawn on if need be, but which can also be returned to its assisted status if necessary. . . . [There will be] a fund of a floating population, of a liminal, infra- or supra-liminal population, in which the assurance mechanism will enable each to live, after a fashion, and to live in such a way that he can always be available for work, if market conditions require it. (Foucault 2008: 206)

We can practically sense the theoretical and political distance of propinquity between Foucault and Marx, especially when we recall Marx's words on the surplus population:

If a surplus population of workers is a necessary product of accumulation or of the development of wealth on a capitalist basis, this surplus population also becomes, conversely, the lever of capitalist accumulation, indeed it becomes a condition for the existence of the capitalist mode of production. It forms a disposable industrial reserve army, which belongs to capital just as absolutely as if the latter had bred it at its own cost. Independently of the limits of the actual increase of population, it creates a mass of human material always ready for exploitation by capital in the interests of capital's own changing valorization requirements.[4] (Marx [1867] 1990: 784)

From this close proximity we can see an opening in Foucault's genealogical project, which is silhouetted within Marx's analysis of capitalist accumulation yet clearly moving away from it, pushing it away while revealing what it is hiding. According to Marx, "It forms a disposable industrial reserve army, which belongs to capital just as absolutely *as if* the latter had bred it at its own cost" (emphasis added). Here we should imagine Foucault's genealogical method revealing how concretely and technically this *as if*—which is the core of capitalist ideology and bourgeois speculative thought—is discursively and materially produced, secured, and naturalized through institutional power and force.

Foucault's method also resonates with Marx's "absolute general law of capitalist accumulation." Marx ([1867] 1990: 798) writes, "The more extensive, finally, the lazarus layers of the working class, and the industrial reserve army, the greater is official pauperism. *This is the absolute general law of capitalist accumulation.* Like all other laws it is modified in its working by many circumstances, the analysis of which does not concern us." Here we cannot avoid hearing a certain impatience behind Marx's narrative voice that interrupts itself, stopping itself from further analyzing these concrete discourses. At the same time, Marx deviates from the consistency of his logical-ontological analysis of capital's inner laws of motion when he speaks of the industrial reserve army. He begins speaking in a unique "classificatory" mode of analysis, distinguishing forms of existence of the surplus populations and diverse strata of the industrial reserve army, which bottoms out in the unclassifiable parts that have no part: the "lumpenproletariat," the lazarus layers. Crucially, Marx draws attention to the presence of a specific discourse that emerges in relation to the most impoverished: *official pauperism.* Not pauperism in itself but pauperism as it is represented in official (i.e., state) discourse. In *Capital,* Marx showed indications of how a more thorough discursive analysis of state and government policies could be carried out (e.g., in his comments on the Poor Laws), although these were still fragmentary historical illustrations of abstract principles. As Marx typically indicated in key passages in *Capital* (e.g., the sale and purchase of labor power in chapter 6*)*, this kind of concrete and discursive analysis, while crucial, nonetheless fell outside the boundaries of his immediate object of inquiry: the great logic of capital. The example of official pauperism clearly indicates this fastidious, methodological boundary-drawing: it is "modified in its working by many circumstances, the analysis of which does not concern us." Foucault, of course, would say that it concerns his genealogical method fundamentally.

A last example of an obscure area between the discourses of Marx and Foucault pertains to the seemingly semantic and academic distinction between Marx's categories of the relative surplus population and the "industrial reserve army." While Marx (and especially Engels, who coined the term) used the terms *relative surplus population* and *industrial reserve army* more or less synonymously, it should be emphasized that they arguably represent the two different methodological levels of analysis discussed above: the logical level of the demonstration of capital's inner laws of motion as an interconnected totality, where the relative surplus population is the logical and contradictory opposite of productive labor, and the second level of state power and policy (e.g., economic policy, social policy, policing in general, and "bloody legislation"), where the logic of capital is distorted on the level of state representation and discourse, and where the abstract relative surplus population is historically distorted and disordered only to be reordered and regimented into a stratified industrial reserve army. Foucault's genealogical method arguably reveals the concrete mechanisms by which an abstract and homogeneous relative surplus population, which exists in opposition to productive labor, is transformed historically and differentiated internally precisely into a reserve army, with all of its connotations of military discipline, chains of hierarchical command, internal processes of differentiation, ranking, policing, and individuating subjects into lifeless statistics. The industrial reserve army is, therefore, an ideological distortion of the relative surplus population.

The above comments on Marx and Foucault are intended to point to a general area in which the external relation of power/knowledge, which is ambiguously related to the logic of capital and to the commodity form in particular, can be given clearer theoretical definition. I will now turn to three points of discussion in Foucault's analysis of biopolitics that give this area more concreteness: the notions of primal war and race war, the problem of scarcity and antiscarcity, and the problem of how the interest of population, as a generality, is secured through legal processes that individuate surplus populations through techniques of competition. All three specify institutional and material techniques that target the most aleatory events of the floating and the surplus populations—keeping in mind that, surrounding Foucault's analysis of power is the structure and the logic of capital, which *overdetermines* networks of power as much as the latter in turn distort and pervert this logic through their own relatively autonomous norms.

Race War and Insufficient Differences

In "Society Must Be Defended," Foucault (2003: 51) speaks of aleatory events occurring within populations in terms of primal war: "We must rediscover the war that is still going on, war with all its accidents and incidents." The accidental nature of the war of which Foucault speaks (in his reading of Hobbes) stems from a condition in which the agents of war are "born in equality" and "in the element of that equality," and thus fight each other within and on condition of, a state of "nondifference or insufficient differences." As Foucault (91) writes, "What happens in this anarchy of minor differences that characterize the state of nature? . . . The absence of natural differences therefore creates uncertainties, risks, hazards, and, therefore, the will to fight on both sides; it is the aleatory element in the primal relationship of forces that creates the state of war." It is on the basis of the accidents and the contingencies of this primal war that Foucault (54) introduces the idea, which will eventually allow him to speak of governmentality, of a growing rationality that is fragile and superficial and that seeks to overcome the aleatory nature of the primal war. Here the growing rationality is organized not only by war but specifically by a race war, which represents the warring milieu in terms of a binary logic of Us versus Them that effects a fragmentation of "the field of the biological that power controls" and that "allows power to treat the population as a mixture of races, or to be more accurate, to treat the species, to subdivide the species it controls, into the subspecies known, precisely, as races" (255). In short, the primal war will produce difference(s) as a way to overcome the unpredictabilities stemming from insufficient differences. This is what Foucault calls the "first function" of racism, which, when combined with the second function of racism— establishing a relationship of war in which "my life" is premised on the "death of the other"—creates, out of these accidents, a fragile, superficial yet growing rationality or governmentality.

In relation to the question of floating populations, the discourse of a race war—which purports to overcome the aleatory element of battles fought in a state of relative equality or "insufficient difference"—becomes meaningful insofar as it represents the masses of surplus populations through the "gaze" of difference and identity. The significance of unemployment, which as Marx says sarcastically, "sets free" workers from industry and into the wider surplus populations, is that in this condition of "liberation" from capitalist production, workers sense an uncanny equality with other workers, for they are are reduced to a condition of dispossession and self-estrangement.

In the condition of unemployment, the commonness of all workers and their potential to work is revealed: structurally, they are in a state of equality of insufficient difference. In the same way that death is said to be a great equalizer among humans, so, too, is unemployment the great equalizer among workers, which creates omnipresent conditions for lazarus-like uncertainty and anxiety in the face of an indifferent God-Money.

What Foucault calls a primal war, which addresses and targets insufficient differences, is the precondition for a racialization of the floating population in state racism and race war. It is a technique of dividing and conquering masses and classes, bifurcating them into binary logics of subrace and superrace. Race war prevents the primal war from developing into a full-blown, class-based war machine against the economic system.

Circulation, Security, and Milieu

In *Security, Territory, Population*, Foucault introduces another way to analyze the aleatory events occurring within a population over time. It is a problem related to circulation, or to the sphere or territory in which circulation can take place. "We see the emergence," Foucault (2007: 93) writes, "of a completely different problem that is no longer that of fixing and demarcating the territory, but of allowing circulations to take place, of controlling them, sifting the good and the bad, ensuring that things are always in movement, constantly moving around, continually going from one point to another, but in such a way that the inherent dangers of this circulation are cancelled out." The dangers of this circulation are then analyzed in terms of how the physiocrats addressed the problem of scarcity of grain. Scarcity of grain led to hoarding by those who possessed it, which led to rises in price and then the possibility of revolt by the masses. The important point of biopolitical power is not simply its defense against the possibility of this revolt but rather its defense against the possibility that this revolt will interrupt the circulation of grain, thereby decreasing profits and national wealth. The antiscarcity system, therefore, comes into existence above all to ensure the smooth circulation (and distribution) of grain and to prevent the possibility of interrupting this circulation. Thus, Foucault showed how thinkers such as Abeille strove toward the creation of an apparatus of security that could ensure endless circulation of the produced commodities, not by intervening directly in the market but rather by intervening in the reality and the manifold variables in nature (e.g., climate, soil, seed, harvesting techniques, and so on), on which the life of grain depends. Foucault (51–60) traces genealogically how these apparatuses of security moved away from a previous obsessive fear of the aleatory political event that could interrupt circulation to a new primary concern with regulating the aleatory nature of the life or reality of the grain itself as a technique of creating a powerful milieu of circulation.

With the rise of nineteenth-century industrial development, in which the most important circulating commodity for the production of profit and national wealth became the peculiar commodity of labor power, this way of thinking about the scarcity of grain increasingly became a paradigm for biopower and its apparatuses of security, which targeted the "natural" life of populations as a site of knowledge production. Here biopolitical power speaks a discourse that asks how to reduce the odds (and preferably eliminate the odds) that these populations ever become scarce. This was done by institutionalizing, Foucault (97) argues, "an apparatus that [would] ensure that the population, which is seen as the source and the root, as it were, of the state's power and wealth, will work properly, in the right place, and on the right objects." This apparatus would then target for intervention, manipulation, and control, not the individual's life (for that is arguably a function of disciplinary power) but rather the elements on which the general life of a population depends for survival and reproduction.

Here, of course, is the familiar place where Foucault's analysis of discourses of sexuality comes into view. At the same time, Foucault (100) says that biopolitical intervention takes place on "a range of factors and elements that seem far removed from the population itself and its immediate behavior, fecundity, and desire to reproduce. For example, one must act on the currency flows that irrigate the country." In other words, a milieu of circulation must be created artificially through these apparatuses of security in relation to the elements on which the life of the population is said to depend, but it is not a form of power that works on the immediacy of somatic bodies. "It is possible to act effectively on the population through the interplay of all these remote factors" (101). A series of mediations, far and near, serve as a mechanism of power that connects the diverse conditions of life of a population into a vast seriality of circuits. Biopolitical power, Foucault says, operates through these apparatuses of security in a "centrifugal" way in which "new elements are constantly being integrated. . . . Security therefore involves organizing, or anyway allowing, the development of ever-wider circuits."[5] The centrifugal movement of this milieu must thus ensure or heighten not simply the biological reproduction of the population but its probabilistic circulation as bearers of the peculiar commodity of labor power, whose possibility of becoming a "scarce resource" must be prevented in a calculating way.[6]

The problems introduced by Foucault on circulation, security, and milieu bring us a step closer to that which Foucault brackets out of his analysis: Marx's analysis of the logic of capital. It is bracketed but in such a way that a historico-theoretical runway is paved toward the encounter with Marx's analysis of the logic of capital. Foucault's analysis of the milieu presents a genealogy of techniques of managing surplus populations, which fell outside of Marx's analysis of the laws of motion of capital. When Marx speaks of how the "absolute general law of accumulation" is modified by many circumstances, these many circumstances have everything to do with what Foucault called the aleatory events that occur within populations over a period of time but also within the area and sphere of circulation. Apparatuses of security not only promote market circulation but also ensure that worker-bodies will abide by the norm of the working day, thus leaving the sphere of circulation and entering (back) into production, thereby contributing toward their reproduction as a species while also realizing surplus value for capital.

Foucault's notion of circulation is thus vastly different from Marx's conception. For Marx, as the opening chapters of *Capital*, volume 1, demonstrate, the sphere of circulation is meaningful as a world in which commodities are exchanged and in which the elementary forms of value in a capitalist commodity economy appear in specific forms: the commodity form and the money form. The sphere of circulation is thus the sphere in which these forms make their appearance as the surface or skin of the capitalist commodity economy. Marx emphatically does not begin his analysis of *Capital*, therefore, with the substance of value (i.e., labor) but with the forms of value that appear in circulation and in the process of exchange. The substance of value, by contrast, is thinkable only on the basis of exchange between owners of money and owners of labor power, who confront each other as owners of commodities before the eyes of the law and thus as individual owners of private property. In other words, Marx outlines the boundaries of his analysis of the logic of the motions of capitalist commodity economies by bracketing out the problem of sovereign law behind the appearance of the commodity form. In the sphere of circulation and in the process of exchange within a capitalist regime, the social form in which parties confront each other is a legally structured one in which bearers of value are free (in the double sense of being free of possessions and free to sell), equal, and motivated by self-interest. Marx thus writes:

> The sphere of circulation or commodity exchange, within whose boundaries the sale and purchase of labor-power goes on, is in fact a very Eden of the innate rights of man. It is the exclusive realm of Freedom, Equality, Property, and Bentham. Freedom, because both buyer and seller of a commodity . . . are determined only by their own free will. . . . Equality, because each enters into relation with the other . . . and they exchange equivalent for equivalent. Property, because each disposes only of what is his own. And Bentham, because each looks only to his own advantage. The only force bringing them together, and putting them into relation with each other, is the selfishness, the gain and the private interest of each. Each pays heed to himself only, and no one worries about the others. And precisely for that reason, either in accordance with the pre-established harmony of things, they all work together to their mutual advantage, for the common weal, and in the common interest. ([1867] 1990: 280)

This passage is the culmination of the first two parts of volume 1 of *Capital*, and it marks a methodological threshold from the sphere of circulation, in which Marx establishes the forms of appearance of value, to the sphere of production, where Marx demonstrates how surplus value is produced on the basis of the consumption of labor power, the consumption of which is only possible on the basis of its sale and purchase in circulation. The point is that Marx methodologically begins his analysis with the appearance of the individual owner of commodities in the sphere of circulation in order to reveal, in his analysis of the process of exploitation and production of surplus value, the antagonistic class contradictions of production that are concealed by the illusory appearance of the commodity.

The uniqueness of Foucault's analysis of biopolitics, by contrast, is that it remains within the sphere of circulation in order to reveal a prehistory behind the appearance of individuals-as-individuals, a genealogical history that culminates in the moment of the sale and the purchase of labor power. From this perspective the sale and the purchase of labor power is the culmination or endpoint of a series of processes and practices that regulate populations, as a generality or species-being, through processes and techniques of racialization and individuation. It is a history of techniques that territorialize a milieu (or area) for circulation and commodification, that install apparatuses of security to probabilistically eliminate random or uncertain elements of (surplus) populations, and that strategically seek to maintain endless circulation as a regulative principle. Harnessing this circulation, however, requires processes of individuation that manipulate the interests of populations. Foucault's analysis of biopolitics thus offers a concrete way to see how the "very Eden of the innate rights of man," which Marx identified sarcastically when speaking of the sphere of circulation, is an effect not simply of juridical law and the work of the sovereign but of infra- or sublegal processes of domination that structure the conditions of living populations.

Interest, Law, and Competition

In *The Birth of Biopolitics*, Foucault (2008: 47) discusses the aleatory events that occur within a population over time in relation to "the utility value of government" that exists within "a regime where exchange determines the value of things." Displacing Marx's analysis of the contradictory character of the commodity form (use value and exchange value), Foucault instead focuses on how government represents what is most useful and effective for constant circulation. Circulation, therefore, is encompassed by and even determines the "utility value of government." Circulation, for Foucault, is not simply a sphere (as it is for Marx) but also, doubly, "a regime where exchange determines the value of things" (Foucault 2004: 44). In short, circulation is a site of not only where value is realized in exchange but also where, through exchange, the limits of public power and the formation of public rights and administrative rights are revealed as an area of extreme fragility and explosiveness. Thus, the utility value of government is a value that emerges in relation to the effectivity in maintaining circulation as a regime of truth. This, Foucault says, is an expression of utilitarianism.

Ultimately, however, what is most useful to and for governments is not only the extent to which circulation can be maintained but how the extensivity of circulation can produce the general interest of the population. This takes place through a series of knowledge-gathering practices and techniques that individuate the multiplicity of populations. Here, for example, the knowledge-collecting technique known as the case makes it appearance: "The notion of case appears, which is not the individual case, but a way of individualizing the collective phenomenon . . . and [of] integrating individual phenomena within a collective field, but in the form of quantification and of the rational and identifiable" (88–89). On the basis of this quantification, rationalization of chance, and the identification of individuals as a method of integrating the latter into a collectivity, the "utility value of government" can grasp what is most "pertinent" to it, namely the general interest of a population, which increasingly becomes an object of knowledge based on probabilistic reasoning and calculations of the variables and elements on which the life of populations is thought to depend (Foucault 2007: 89). Most simply put, the government is that which manipulates these interests. In this manner, Foucault discovers a logic of strategy, one that supplements (Marxism's) dialectical logic (Foucault 2004: 46, 44).

How do processes of individuation take place in such a way that the interest of populations is produced for a government of utility to know and regulate in a regime of exchange? It is around this question of strategy that Foucault discusses three interrelated problems: law, competition, and social policy. To oversimplify the interrelation greatly, we could say that, to protect and maintain circulation, government must step back, as it were, from direct economic intervention, but only in order to step up, increase, and intensify legal practices on the social conditions of economic movement. The question is not how governments should intervene in economic movement but how governments should intervene in the conditions of economic movement. Through the form of law, through the rule of law, and through a whole panoply of legal techniques that ultimately represent the principle of private property, biopolitical power zeroes in on the conditions of economy by individuating populations through legal means, as if a legal screen filters populations into individuated elements, thereby creating a highly policed milieu of economic competition, which is believed (ideologically) to ensure smooth circulation. In this way biopolitical power strategically individuates populations into legal subjects or identities as a way to harness the general interest of the population. Lastly, social policy becomes the primary instrument by which, "concrete and real spaces in which the formal structure of competition [can] function" (Foucault 2008: 131–32). Competition becomes not a "positive theory, but . . . a normative theory, an ideal type that one must strive to achieve" (128 n54). The real effect or function of social policy ultimately is thus not to provide social care or security (for that would be its ideological fantasy, we should say) but rather to put into action this process of individuation through various legal channels and techniques as a way to carve out an economic space in which these individuals must dwell competitively. "It involves," Foucault (44) writes, "an individualization of social policy and individualization through social policy. . . . In short, it does not involve providing individuals with a social cover for risks, but according everyone a sort of economic space within which they can take on and confront risks."

Biopolitics, Imperialism and Bolshevism:
The Example of Unemployment in Japan

Before returning to this notion of producing an economic space in which these individuals can take on and confront risks, I would like to briefly give a concrete example of how the biopolitical problems of law, competition, and social policy can be seen in action in interwar Japan. Here I take up the specific example of the Unemployment Emergency Relief Program (UERP, or 失業救済事業) and look at how the work of the UERP had an effectivity within the sphere of circulation (see Kawashima 2009: 169–203). This effectivity was closely related to the problems of race war, racism, and the bifurcation of surplus populations according to ethnicity. These problems became enormously relevant for an understanding of capitalism in its imperialist stage, particularly in light of the Bolshevik Revolution of 1917.

Created in the winter of 1925 in the crucible of the post–World War I recession in Japan, and under the banner of providing social relief to all unemployed masses in Japan, the UERP system was, almost immediately after its institutional inception, overwhelmed by enormous waves of masses in immediate need of employment in order to survive. Underfinanced and underequipped to meet all of their needs and demands, the UERP faced an explosive social and political situation on a daily basis. What emerged institutionally was not merely a series of stopgap measures to defuse a potentially explosive situation. Rather, a series of regulations emerged—one that was backed by law, various court systems, and the police—that affected a general interest of the surplus populations to register with the UERP system as a condition for selling their labor power as a commodity in order to obtain wages with which to consume their daily necessities. How was this general interest achieved? There were three particularly clear ways.

First, we have a combined effect of the work of the law and the process of individuation. The UERP had the lawful authority to require all workers wishing for unemployment assistance and aid (in the form of labor introductions to sites of work) to register with the UERP. The UERP, however, only recognized workers registering as individuals with identifiable identities as well as with identifiable residential addresses, which structurally discriminated against homeless and migrating populations. If the workers could not be identified positively, registration to the UERP was denied, leaving the worker to find assistance in other ways. The overall effect of the legal process of registration, therefore, was one of individuating the flows of surplus populations and of creating a peculiar self-interest within the collective masses of workers, to be identified as individual residents, in order to become recognized by the UERP. Institutional recognition became an absolute precondition for work.

Second, and as a way to understand the utility value of the government in relation to the logic of race war, the UERP also was the site of a process of bifurcating the surplus populations according to ethnicity, and specifically according to whether the workers were Japanese or Korean. The catalyst for this bifurcating process was that between 1925 and 1928, unemployed Japanese workers vociferously complained that they were unable to register with the UERP because the presence of large numbers of Korean workers in the UERP offices crowded out Japanese workers. In response to this, in 1929 the UERP created a system whereby Japanese workers wishing to register with the UERP were given preferential treatment over Korean workers, even though the UERP's officially stated pledge was to be open to all workers in the Japanese Empire. This racist, preferential treatment was achieved by creating an additional legal dimension to the process of registration. This took the form of the aleatory introduction of mandatory workbooks, or *rōdōtechō*, that workers were suddenly required to obtain as a legal precondition for registering with the UERP. In this sense it further solidified the will of workers to register as individuals. At the same time, however, a bifurcation of the unemployed workers along ethnic codes was instituted by the UERP by differentiating the techniques and channels through which workbooks were distributed. Japanese workers could obtain the workbooks from police stations, the UERP offices themselves, or district committee offices called *hōmen i'inkai*. Korean workers, however, were denied workbooks at these sites and were required instead to obtain them at private or semipublic welfare organizations specializing in Koreans, such as the notorious *Sōaikai*, or at publicly operated Korean welfare organizations, such as the *Naisen Kyōwakai* in Osaka. The point is that the logic of race war—logic that tended to naturalize notions of Japanese as a superior race over Koreans—was institutionalized through techniques of distributing workbooks that created a bifurcation of the floating population along ethnic codes, which had the strategically planned effect of dividing and separating class-based solidarities between Japanese and Korean workers.

Third and last, the UERP artificially created a milieu for competition to emerge within racialized and individuated elements of the floating population. This was the final and most insidious technique that created and harnessed a general interest of the floating population and that also was related to the distribution and possession of the workbooks. As mentioned above, Koreans could only obtain workbooks at places such as the *Sōaikai*. But Koreans were required to register and become members with these organizations as conditions to receive the workbooks. A problem, however, is that many Koreans despised organizations such as the *Sōaikai* because of their antilabor union politics and close connections to the Japanese police. On the other hand, other Koreans agreed to become members, often simply as a desperate means to register with the UERP. The overall effect was a bitter series of jealous struggles among and between Koreans over the possession of workbooks. Many hoarded them, others sold or stole them, while others borrowed or lent them, leading the UERP to institute a process by which fraudulent uses of workbooks could be identified. Thus, after 1929, the affixing of photographs of registered workers on the workbooks became standard practice. This revealed the state's intolerance of unidentifiable and therefore unindividuated, anonymous surplus populations. At the same time, the workbooks became fetish objects whose possession was prized and jealously fought over.

As an extensive bureaucratic machine that combined intensive processes of individuation, racialization, and competition through the rule of law, the UERP institutionalized the problem of what Foucault called "the rule of non-exclusion." "In the idea of an economic game," Foucault (2008: 202) writes, "we find that no one originally insisted on being part of the economic game and consequently it is up to society and to the rules of the game imposed by the state to ensure that no one is excluded from this game in which he is caught up without ever having explicitly wished to take part." The force of the UERP was not simply that it excluded certain workers from the processes of unemployment aid because of their ethnicity. Korean workers were never simply excluded but were integrated (and excluded) strategically, selectively, and differentially into labor activity. This was achieved by legal means that individuated surplus populations in an artificially produced yet real milieu of competition revolving around the distribution and possession of UERP workbooks. Through various apparatuses of security, the UERP thus mobilized a general interest of floating populations by a principle of nonexclusion, which instituted technical and legal processes of individuating (racialized) subjects and accorded them a space within a regime where exchange—which was premised on a structure of competition—determineed value as truths.

The example of the UERP in interwar Japan also raises, first, the problem of biopolitics in light of what Lenin called the highest stage of capitalism, that is, the stage of imperialism, and second, the Bolshevik Revolution of 1917. Foucault only seems to have flirted with how an analysis of biopolitical power could be carried out specifically in the stage of imperialism (e.g., his analysis of Nazism). One reason, we can surmise, is that Foucault believed his analysis of biopolitical power—an analysis of concrete forms and relations of domination and oppression—could also be used for an analysis of socialist regimes as much as for capitalist ones. Nonetheless, it is clear that the problem of—or better, the word—*imperialism* (similar to other words like *commodity*) is one that Foucault chose to bracket out of his analysis for his own strategic reasons. However, it also is clear that in our contemporary conjuncture in which post-Soviet, post–Cold War ideologies of capitalist triumphalism are rampant, this kind of bracketing is not only unnecessary but also an obstacle to producing more specific and concrete analyses of the always already-passing present that include salient characteristics of imperialism that are not immediately reducible to the genealogy of biopolitical power but that, for this very reason, present another relation of nonrelation to the problem of biopolitical power.[7]

To be consistent with the overall scope of this essay, I would like to mention some of the salient characteristics of imperialism from the perspective of the problem of floating populations. Following Lenin's analysis, once the capitalist commodity economy entered the stage of imperialism; once, that is, the concentration of capital led to forms of monopolization in industry, leading to monopolization in the banking system; once, in short, the capitalist commodity economy became dominated by finance capital and the export of capital as the condition for colonialization and "the carving up of the world," the problem of floating populations not only became a global problem of the international division of labor, it also became a chronically existing phenomenon. How is this chronic existence to be understood? We can mention at least two factors. First, this chronic existence can be understood as an effect of the way governments and corporations in the stage of imperialism are afforded the power through the use of finance capital and extensive mechanisms of credit to circumscribe territories for the movement of capital without eradicating noncapitalist populations (e.g., so-called feudal remnants), choosing instead to maintain them as an underdeveloped surplus population, thereby contributing to the chronic existence of a surplus population on a worldwide basis. In other words, this is a kind of imperialist integration through sustained and strategic underdevelopment (especially in the colonies) that does not so much treat so-called feudal remnants as obstacles to capitalist development but instead integrates the social and cultural remnants of older modes of social organization as a means to further depress overall wage levels, which reflects an increasingly expansive surplus population.

Second, the chronic existence of floating populations in the stage of imperialism is a symptom of how phases of recession, under the dominance of finance capital in the stage of imperialism, take on a chronic character while phases of economic prosperity become increasingly shorter and shorter. Moreover, the phase of economic crisis itself usually appears at lightning speed affecting all sectors of the economy, often seemingly at the same time. Compared to the previous stage of liberal capitalism (or laissez-faire), economic cycles in the stage of imperialism no longer appear as phenomena that abide by a clear periodicity (e.g., decennial cycles). Cycles now appear unpredictably and erratically, with phases of recession dragged out, thus contributing to the chronic character of surplus populations. As surplus populations increasingly took on a chronic character within capitalist commodity economies, the problem of governing populations became more and more urgent as a defensive means against social revolution and to maintain the circulation of commodities and money for the accumulation of capital.

It is here, of course, that the Bolshevik Revolution of 1917 is extremely significant, particularly in the ways capitalist nation-states attempted to govern these populations through social policies and techniques of policing. In the wake of the Bolshevik Revolution the chronic character of surplus populations could not but appear in an overdetermined and political way that embodied an actual potential to transform the chronic quality of the existence of surplus populations into a conscious and organized refusal to integrate into the circuits of capital that produced them in excess of capital in the first place. After the event of Bolshevism, the semiotic valence of the sign of unemployment could not avoid appearing as a revolutionary sign of an imminent overcoming of the irresolvable class contradictions of the capitalist system. Social policies, especially in the capitalist regimes, were thus put into action partially, if not wholly, to stem the tide of Bolshevism in an era when chronically existing surplus populations increasingly embodied and expressed this social contradiction in political and revolutionary terms.

It seems to me that this political and ideological dimension of social and economic policy in the capitalist regimes after 1917 both influenced and were influenced by discourses on biopolitics. In short, we can see how counter- or antirevolutionary ideologies impacted within capitalist social and economic policy after 1917 could easily take up and put into institutional action notions of race war that biopolitical discourse kept alive in relation to other related signs such as hygiene, insurance, demography, and sexuality, and so forth. What begs emphasis here, however, is the close relationship of biopolitical discourse to ideologies of society that evacuate the problem of class struggle through signs of individual interest, race, and ethnicity. In the case of fascist Japan, for example, it is difficult if not impossible to understand the ideas of the "Greater East Asian Co-Prosperity Sphere" apart from the soil of liberal ideology that developed during the earlier Taisho period and through it passed (as Tosaka Jun would point out). Liberal ideology facilitated the ideologies of Emperor worship and Japanism on the basis of an already sacralized discursive production of the individual-as-subject, notions of society in terms of pure ethnicity, both of which worked to discursively disavow the reality of class struggles tearing apart Japanese society and the Japanese Empire. This ideology, however, was enunciated through various biopolitical discourses on race, ethnicity, nation, and multiethnicity. Especially after 1917 in the capitalist world, it is impossible to ignore the problem of ideology, especially when analyzing problems of biopolitical discourse. Biopolitical discourse racializes the international division of labor and artificially creates—out of a single mass of (surplus) populations—a subrace and a superrace. This points not only to the problem of discourse but also to the problem of an ideological fantasy that disavows and represses the material reality of class struggles.

The World of Commodities beyond Biopolitics

If Foucault's analysis of biopolitics takes us to the relationship between the utility value of government and the regime of exchange, and if his analysis of apparatuses of security, techniques of competition, and processes of individuation show us how biopolitical power—via social policy—does not so much protect individuals from risks as it accords individuals an economic space within which they take on and confront risks, then what requires further theorization is the precise nature of the relationship of biopolitics to the regime of exchange, which institutionally fixes and normalizes value and truth. However, how is the truth of the exchange process in circulation related to the space carved out by biopolitical power in which individuals are forced—by the principle of nonexclusion—to experience and confront risks? What is the meaning of this risk?

Here Marxism must take a clear stance on this issue, for we are not dealing primarily with the problem of risk and probability on the biological level of the species in relation to, say, disease. Nor are we simply dealing with the angst-filled risks found in Heidegger's "being toward death." Nor is it even simply the risk of selling commodities in speculative trading, financial gambling, and so on. Rather, we are talking about the specific risk that structurally inheres in the position of selling commodities: not just any commodity but the peculiar commodity of labor power, which exists—latently and polymorphously—in the form of surplus populations. The problem of biopolitical power leads us to think and analyze, as a necessary outcome of Foucault's analysis, the aleatory event(s) of the commodification of labor power. It points to the arduous and often torturous road traversed by the expropriated and dispossessed masses that leads unpredictably towards changing norms of the working day of the capitalist mode of production—a mode of production that exists alongside its attendant "forms of intercourse" (Marx). Biopolitical discourse additionally exists alongside these forms of intercourse (exchange), delineating strategic areas for a "regime of veridiction." Overall, Foucault's genealogy of power reveals the tenuous and fragile historical connections between what Marx calls the spheres of circulation and production and between expropriation and exploitation. Between them is a milieu and an area where excessive force is deployed strategically as surplus power.

In this light, the usefulness of Foucault's analysis of power is the way in which he allows us to see how subjects are produced as individuals within the sphere of circulation, that is, on this side of the commodification of labor power before a sale has been made, as it were. In this regard Foucault's analysis speaks to Marx's chapter in *Capital* titled "The Sale and Purchase of Labor Power," a pivotal chapter in *Capital, Volume 1,* insofar as it is the chapter that ends Marx's analysis of the abstract forms of circulation of the capitalist commodity economy that began with the commodity form. It also inaugurates part 2 of the book, which focuses on production. In chapter 6, Marx writes as if he has not already theoretically worked out how the capitalist mode of production eventually will be shown to create its own conditions of reproduction by producing surplus populations as a result of methods of extracting surplus value. The key condition of reproduction, of course, is the separation of owners of money and owners of labor power. This separation is the *sine qua non* of the commodification of labor power in which owners of money (i.e., the capitalist in Marx's analysis) must be lucky enough to encounter bearers of labor power in the sphere of circulation.

Let us consider the following passage by Marx. If we read this passage in its entirety, we can see more clearly where Foucault's analysis of biopolitical power supplements Marx's self-proclaimed limits of his analysis of exchange:

> In order to extract value out of the consumption of a commodity, our friend the money-owner must be lucky enough to find within the sphere of circulation, or the market, a commodity whose use-value possesses the peculiar property of being a source of value, whose actual consumption is therefore itself an objectification of labor, hence a creation of value. The possessor of money does find such a special commodity on the market: the capacity for labor, in other words, labor power. . . .
>
> Why this free worker confronts him in the sphere of circulation is a question which does not interest the owner of money, for he finds the labor-market in existence as a particular branch of the commodity-market. *And for the present it interests us just as little. We confine ourselves to the fact theoretically, as he does practically.* One thing, however, is clear: nature does not produce on the one hand owners of money or commodities, and on the other hand men possessing nothing but their own labor power. This relation has no basis in natural history, nor does it have a social basis common to all periods of human history. It is clearly the result of a past historical development, the product of many economic revolutions, of the extinction of a whole series of older formations of social production. (Marx 1990: 270, 273; my emphasis)

Within Marx's analysis of capitalist production and accumulation, the process of expropriation—discovered in so-called primitive accumulation—is hidden behind the signs of capital, including the relative surplus population that capital produces as a result of its own process and as a prosthetic of the originally absent labor power commodity. The secret that Marx exposes is that expropriation is the repressed precondition for exploitation. However, from Foucault's perspective, located at the intersection of the utility value of government and the regime of exchange, the process of expropriation does not lead inevitably to exploitation. Rather, the work of biopolitical and disciplinary power as a combined logic of strategy governs the material conditions in which the expropriated are divided, classified, and individuated in anticipation of their future exploitation in production—that is, in anticipation of their entry into capital's logic of the dialectic.

In this regard the analysis of biopolitical power is an important supplement and complement to Marx's analysis of the law of populations peculiar to capitalist commodity economy and to his critical analysis of the state's sovereign power. The Foucauldian question of power, strategy, and truth exists between Marx's analysis of capital's law of populations and the analysis of the state's sovereign laws. For our purposes it asks how does the bearer of labor power, while still in the formless and anonymous multiciplicity of a surplus population, become individuated as a subject in possession of labor power—one who is willing, moreover, to daily occupy the position of an existential risk in selling it. Echoing Marx, Foucault's analysis reveals how the relationship of exchange is hardly a natural, inevitable, or necessary relation but one that has been strategically forced into being.

In conclusion, where the revolutionary methods of Marx and Foucault intersect (albeit on starkly different levels of analysis) is arguably around the problem of the commodification of labor power. This is a process that is presupposed in Marx's analysis of the laws of motion of capital but that also is unexplored in Foucault's studies of biopolitical power. One of the most distinctive aspects of the process of commodification is that it discloses an unavoidable risk and contingency of selling commodities: capitalism's aleatory event par excellence. We can now add that the problem of the aleatory, which is endemic to exchange, also emerges—doubly—from within the sphere of circulation in the form of biopolitical and disciplinary power.

Biopolitical discourses of power/knowledge fill in the void between capitalist circulation and production; it is the soil out of which capitalist ideology takes root.

The stakes of considering the commodification of labor power as a problem that mediates the methods between Marx and Foucault is that we can truly start to analyze how the aleatory nature of networks of power lead us to consider the aleatory nature of the process of exchange on the basis of which the capitalist commodity economy is instituted and in relation to which the networks of power find their functional raison d'être. What is the historical nature of this passage from the aleatory nature of power to the aleatory nature of commodification? Is there not found here an unrepresentable splitting—a clinamen—distinguishing opposing classes and their mutually repelling, peculiar fears: those who must sell labor power in the face of nothingness and emptiness (the proletariat) and those capitalists who, in the last instance, fear the dreadful thought that labor power might be permanently absent from the working day? How do these contradictory fears shake and tremble alongside the sheer paranoia and narcissism of networks of power? (Is there identity?!) If networks of power are interconnected to and with networks of commodification, is not one of the most urgent tasks today to produce interconnected, *countermilieus* to these networks that bring about their aleatory nature in a permanent fact that brings about a practical end to ideologies of capital's necessity? It is with these interconnections in mind that we can remind ourselves of Marx's inspiring statement about the revolutionary potential of an aleatory event that exists already—virtually in the theoretical analysis of the intrinsic interconnections of the capitalist mode of production. In a letter to Kugelman from July 11, 1868, Marx wrote, "When faced with the disclosure of the intrinsic interconnection [of the capitalist mode of production], the vulgar economist . . . clings to its appearance and believes them to be the ultimate. Why then have science at all? But there is also something else behind it. Once interconnection has been revealed, all theoretical belief in the perpetual necessity of the existing conditions collapses, even before the collapse takes place in practice" (Marx [1868]).

Notes

1 Karatani's *Marukusu sono kanosei no chushin* (Marx: The Center of His Possibility) was originally published in 1973. This important text rereads Marx's law of value from the perspective of the position of selling, a theme repeated and expanded upon in many of his subsequent works. In English, see also his *Architecture as Metaphor: Language, Number, Money,* Cambridge: MIT Press, 1995.

2 See the chapter titled "The Underground Current of the Materialism of the Encounter" in Althusser, 2006.

3 On labor power and primitive accumulation, see Gavin Walker, *The Sublime Perversion of Capital*, chapters 3 and 4, Durham, NC: Duke University Press, 2016.

4 Marx ([1867] 1990: 794) defines the "surplus population . . . in the floating form" as "the workers [who] are sometimes repelled, sometimes attracted again in greater masses, so that the number of those employed increases on the whole, although in a constantly decreasing proportion to the scale of production. Here the surplus population exists in the floating form."

5 By contrast, disciplinary power, Foucault says, operates in a "centripetal" way that tends toward concentrating and enclosing spaces.

6 For Foucault on the notion of the milieu, see Foucault 2007, 35–36: "The space in which a series of uncertain elements unfold is, I think, roughly what one can call the milieu. . . . What is the milieu? It is what is needed to account for action at a distance of one body on another. It is therefore the medium of an action and the element in which it circulates. It is therefore the problem of circulation and causality that is at stake in this notion of milieu."

7 At least in the lectures on biopolitics, Foucault (2008: 56) avoided explicit treatments of the topic of imperialism, stating only on occasion, for example, that the birth of biopolitics was not reducible either to colonization or to "the start of imperialism in the modern or contemporary sense of the term, for we probably see the formation of this new imperialism later in the nineteenth century." Foucault rather chose to trace how biopolitical power, especially in its eighteenth-century manifestation at the end of the stage of mercantilism, revealed "a new type of global calculation in European governmental practice" and "a new calculation on the scale of the world" (56).

References

Althusser, Louis. 2006. *Philosophy of the Encounter: Later Writings, 1978–1987*, translated by G. M. Goshgarian. London: Verso.

Foucault, Michel. 1981. *Remarks on Marx: Conversations with Duccio Trombadori*, translated by James Cascaito and R. James Goldstein. New York: Semiotext(e).

Foucault, Michel. 2003. *"Society Must Be Defended": Lectures at the Collège de France, 1975–1976*, translated by David Macey. London: Palgrave.

Foucault, Michel. 2004. *Naissance de la biopolitique: Cours du Collège de France, 1978–79.* Paris: Gallimard.

Foucault, Michel. 2007. *Security, Territory, Population: Lectures at the Collège de France, 1977–1978*, translated by Graham Burchell. London: Palgrave.

Foucault, Michel. 2008. *The Birth of Biopolitics: Lectures at the Collège de France, 1978–1979*, translated by Graham Burchell. London: Palgrave.

Karatani, Kojin. (1973) 1985. *Marukusu sono kanosei no chushin* (*Marx: The Center of His Possibility*). Tokyo: Kodansha.

Karatani, Kojin. 1995. *Architecture as Metaphor: Language, Number, Money*, translated by Sabu Kohso. Cambridge, MA: MIT Press.

Kawashima, Ken C. 2005. "Capital's Dice-Box Shaking: The Contingent Commodifications of Labor Power." *Rethinking Marxism* 17, no. 4: 609–26.

Kawashima, Ken C. 2009. *The Proletarian Gamble: Korean Workers in Interwar Japan*. Durham, NC: Duke University Press.

Marx, Karl. [1868]. "Letter to Kugelmann, July 11, 1868." Marxist Internet Archive. www.marxists.org/archive/marx/works/1868/letters/68_07_11-abs.htm (accessed February 7, 2017).

Marx, Karl. [1867] 1990. *Capital*. Vol. 1. New York: Penguin Classics.

Marx, Karl. [1894] 1991. *Capital*. Vol. 3. New York: Penguin Classics.

Walker, Gavin. 2016. *The Sublime Perversion of Capital: Marxist Theory and the Politics of History in Modern Japan*. Durham, NC: Duke University Press.

Forces and Forms:

Governmentality and *Bios* in the Time of Global Capital

Sandro Mezzadra
Translated from the Italian by Gavin Walker

In the World

What matters for "biopolitical investigations," writes Roberto Esposito (2006: 7) in a recent article devoted to the "History of Concepts and the Ontology of Actuality," "is not the relationship between before and after, but the one between inside and outside. This is how an 'outside'—in this case life—enters into something else (something that is thought to be other) and transforms, twists, or undermines it. What is at stake are not the forms, but the forces [*le forze*], that is, the way in which these forces are exploding the forms." In this brief contribution, I would like to again take up Esposito's point, albeit in a different context from the highly philosophical one in which his thought operates. The tense relation between inside and outside (between forces and forms) will be analyzed here with specific reference to the way in which "life itself," in a sense quite different from its general

positions 27:1   DOI 10.1215/10679847-7251858

meaning, is increasingly one of the most critical sites throughout which the processes of valorization in contemporary capitalism are spreading.[1]

To speak of contemporary capitalism, on the other hand, is to speak, again not in a merely general way, of *global* capitalism: a capitalism, in other words, that seems to have swept away all spatial limits to its expansion, making a simple definition of terms such as *inside* and *outside* problematic, to say the least. Here is an initial question worth emphasizing. One of the fundamental problems we face today is that critical thought needs to forge concepts that are able to grasp the emergence of the *world* as a completely material dimension in which increasingly the most local of our experiences are themselves located. As a large part of the literature on globalization is coming to recognize, it is a world that is at the same time *increasingly unified and increasingly divided*, in which the operation of vectors of homogenization and connection are determined by the continuous multiplication of the factors of "diversity" [«*eterogeneità*»] and the activation of logics of disconnection and fragmentation.[2] It is far from a "smooth" world, and thus we cannot attempt to think it critically through "smooth" concepts. On the contrary, we need "wrinkled" concepts, concepts that are sufficiently flexible to be applied on different scales, capable of grasping simultaneously the elements of unification and division (homogeneity and heterogeneity) to which I referred earlier. What we truly need are concepts that can be applied to the *world*.

The biopolitical is undoubtedly a concept that seems a natural fit for this role—first, because "*bios*," like "world" refers to something we all *have in common*. Thus it seems all the more questionable to me that a substantial part of the contemporary debate on biopolitics takes place within a spatial bordering [*perimetrazione*] roughly covering the term *West*, a bordering that is continuously being put into tension by the set of processes we are experiencing today. In this case we ought to at least keep in mind what Paul Ricoeur has referred to, in a completely separate context, as the hermeneutics of suspicion. This is the case in the example of the recent book by Nikolas Rose, *The Politics of Life Itself.* Let me be clear: this is an important book, containing provocations that it would be neither possible nor desirable to simply dismiss out of hand. I refer, for example, to the emphasis the author places on the need to examine the laboratories in which contemporary biologists work as "a kind of factory" that produces "new forms of molecular life" and "new ways of understanding life"—in which new standards are emerging in relation to which critical thought must redefine its toolbox rather than merely lingering in the position of the "rearguard" (see Rose 2007). But the obstinate limitation of Rose's analysis to the "rich West" (81, 176), as will be seen later, far from being the necessary and given object of analysis, in my opinion, substantially limits its critical impact.

Governmentality and Biopower

I have not mentioned Rose's work by accident. Quite apart from his latest book, this work has in fact been quite influential. As it is well known, along with the work of Paul Rabinow, it has oriented the reception of Foucault in the Anglo-Saxon world in a decisive way, especially regarding the uses and meanings of the concept of *governmentality*.[3] In a paper written by Rose together with Rabinow, the operation of contemporary biopower is presented as the completed paradigm of a postsovereign synthesis of technologies of the self and technologies of government, with liberal concepts of autonomy and the spirit of enterprise. Critical in their approach to the work of Giorgio Agamben, paradoxically, Rabinow and Rose in reality confirm his thesis, according to which sovereignty operates solely through the logic of the exception. Yet, according to Agamben, this logic, which is now widely disseminated, rather appears in their eyes to be merely residual, confined to the Iraqi inferno, to Guantanamo Bay, and eventually to the detention camps for migrants, which are in truth fairly common in Europe and elsewhere. The horizontal and "soft" [«*dolce*»] or "pastoral" matrix of neoliberal governmentality would instead consist in something different from the latter (see Rabinow and Rose 2006).[4]

Obviously, these are themes that return in Rose's most recent book. Informed consent, autonomy, voluntary action, and choice seem to define the very concept of governmentality in his view, even beyond the field of biomedicine, to which *The Politics of Life Itself* is dedicated. In this way, he not only tends to blur the line between ethics and power but, moreover, as Rose himself explicitly states, "between coercion and consent" (Rose 2007: 74). It seems to me that there is something that simply does not work in this theory of governmentality, which ends up negating the persistent role played by *violence* in the neoliberal regimes of the biopolitical and the bioeconomic (Bazzicalupo 2006a; Mbembe 2003), and that celebrates the sovereignty of a given model of the individual (specifically the liberal individual), simultaneously refusing to consider the continuing reproduction—as the material condition of this sovereignty itself—of other "forms of life," other models of subjectivity, in which the distinction between consent and coercion is far from invisible (see Papadopolous and Stephenson 2006; esp. 10). The obstinate limitation to the "West" of Rose's analysis thus appears as a symptom of a more general problem, one that suffuses the entire edifice of the concepts that he utilizes in thinking the biopolitical.

However, it is not, in my opinion, necessary to move far from England (in the case of Rose) to understand the relevance of these problems. But clearly a certain intervention into the "great and terrible world," in the words of Gramsci, would help to clarify the issues. This is a question which it will be necessary to explore also under the vantage point of historiography, by moving toward the necessity of radically questioning the deafening silence of Foucault on the colonial laboratories in which the essential characteristics of modern biopolitics were defined. In fact, it was precisely the colonial laboratory that from the beginning sharply demonstrated the contradiction between the liberal emphasis on the need for knowledge of self and individual self-government on the one hand, and the ubiquity of violence and terror on the other (see on this point Legg 2007: esp. 150).[5] But let us remain for the time being in our present: in particular, studies of fundamental importance for the discussion of biopolitics and governmentality having been emerging from India for many years now—we should assign these a central role that so far, especially in the Italian debate, was denied to them.

Democracy in India

It's sometimes said that if a young Tocqueville wanted to understand what democracy is at the beginning of the twenty-first century, he should go to India rather than America. In India he would be faced with a democratic process that is entirely rooted in the life of *populations*, that is traversed by powerful tensions and yet manages, in the everyday, to produce democratic formulas of composition within these tensions.[6] Whether or not these formulas will prove effective in the medium and long term is a question that is too complex to be merely put forward here in the present essay. But it is clear that today India is one of the most extraordinary laboratories in which forms of experimentation related to the concept of democracy are being undertaken (we will see later, moreover, that this is not the only sense relevant to our discussion of the "laboratories" and "experimentation" in contemporary India).

An integral part of this experimentation consists in the emergence of a "popular" and "subaltern" space of politics, produced and reproduced by an infinite number of daily clashes and negotiations, irreducible to the classical logic of representative democracy, or to the figure of civil society and the language of rights, but one that instead exhibits significant analogies with the history of the biopolitical governmentality of populations. Partha Chatterjee (2006) has proposed to utilize the concept of "political society" to define this space. Let us leave aside here any consideration of the torsion to which this concept becomes thus subordinated with respect to the original Gramscian coinage of the term. Let us also avoid taking up the critical observations that seem more relevant and that I am essentially in agreement with, by Ranabir Samaddar, who points out that the space of political society ends up appearing in Chatterjee to be completely saturated by the actions of governmental power, and barely seems to admit the existence of the continuous autonomous *action* of the "subaltern" (see Samaddar 2007: 1:107–37). What seems important to me is Chatterjee's identification—as a characteristic feature of Indian democracy—of an entirely *biopolitical* dimension, in which what is at stake (within the clashes and negotiations previously mentioned) is at once the material reproduction of the life of the immense mass of "populations."

With this aspect in mind, we must briefly discuss the work of an Indian economist, Kalyan Sanyal, who in a recent book has, in a certain sense, offered a materialist foundation to the concept of "political society" introduced by Chatterjee. Through a fascinating reconstruction of the debates on development that took place after the end of World War II, and in particular through a meticulous analysis of the planning policies adopted in the newly independent India by Jawaharlal Nehru, Sanyal comes to demonstrate, in a manner that I find quite convincing, that one of the most important characteristics of "postcolonial" capitalist development consists in the continual reproduction of the conditions of *primitive accumulation*. This form of primitive accumulation—far from being limited merely to the moment of industrial "take off" characteristic of the model of modernization theory—constitutes a structural element of postcolonial capitalist development, and in the manner classically described by Marx in chapter 24 of the first volume of *Capital*, continually reproduces a "wasteland of the dispossessed" who exceed capitalism's actual need for labor power, and who are not intended to be "integrated" into the process of generalization of wage labor, understood as the privileged means of access to full political and social citizenship (Sanyal 2007).

The most original analysis of Sanyal, however, consists in his insistence on the fact that for a number of reasons—including the essential role played by the heritage of the anticolonial struggles in the origin of Indian independence—in contrast to what happened in the origins of modernity in England, capital in India cannot let the dispossessed die, but rather capital here is somehow forced to take charge of their *life*. The *legitimation* of capitalism thus occurs—in India as elsewhere in the postcolonial world—as a set of processes aimed at ensuring the reproduction of a subsistence economy; it is, in other words, the economic side of the "political society" discussed by Chatterjee. It is not only public agencies that play an essential role in these processes—in addition, of course, to the extraordinary creativity of the "dispossessed" themselves—but also international organizations and nongovernmental organizations that contribute to defining a specific regime of biopolitical governmentality (see Sanyal 2007: chap. 4).

In the Laboratories of Biocapital

I have discussed Sanyal's book elsewhere in greater detail, emphasizing—in addition to its great importance—certain points that seem to me to be some of its basic limits, relating in particular to his argument that the spaces of the subsistence economy within *capitalism* should be considered external to *capital* (see introduction to Mezzadra 2008). However, what should be emphasized here is the impossibility of reconciling the regime of governmentality described by authors such as Sanyal and Chatterjee, a regime in which violence and the specter of death are structural presences notwithstanding the emphasis this regime places on the need to ensure the reproduction of the *life* of the dispossessed, with the "paradigm" proposed by Rabinow and Rose. Further, it would be worthwhile to trace the diffusion of this regime also throughout the cities of Europe and North America, particularly when it is a question of the governing of the movement of migrants or—as in the emblematic case of the French *banlieues*—populations "of immigrant origin" (see Chatterjee 2006).

Sanyal's reference to the persistence of "primitive accumulation" (or more accurately, the *so-called* primitive accumulation, to follow the Marxian formulation) places it in dialogue with a particularly interesting set of recent developments of critical thought derived from unorthodox Marxism.[7] This allows us to reassess—from a rather specific point of view—the themes of biomedicine, so important for the contemporary debate on biopolitics and bioeconomy. In an extraordinary "multisited" ethnographic study, conducted between the Bay Area, the Indian state of Andhra Pradesh and the boundless urban territory of Mumbai, Kaushik Sunder Rajan (2006) has given us an essay bearing witness to the relevance of the issues related to primitive accumulation in order to understand the functioning of "biocapital": namely, the set of social relations that are rooted in the valorization processes of the capital invested in biotechnology, in "postgenomic" medicine, and in the tendential development of personalized biomedicine.

One of the "sites" in which Rajan conducted his research was the Wellspring Hospital, a hospital opened in Parel, right in the heart of Mumbai, by an Indian pharmaceutical company, Nicholas Piramal India Limited (NPIL). The Wellspring Hospital hosts within its own structure Genomed, a start-up launched jointly by NPIL and the Center for Biomedical Technology (CBT), a genomics laboratory that is the pride and joy of Indian public research. Genomed runs clinical trials on behalf of large "Western" pharmaceutical and biotech companies, making available not only the great capabilities shaped through public investment in the research sector in India but also and most importantly, a *population*—the Indian population, in fact—that reproduces within it the entire internal spectrum of global genetic diversity. As S. K. Brahmachari, director of the CBT and a member of the Administrative Board of Genomed emphasizes: "If they want Caucasians, we'll give them Caucasians; if they want Negroids, we'll give them Negroids; if they want Mongoloids, we'll give them Mongoloids" (cited in Rajan 2006: 95).

It is clear that the words of Brahmachari allude to a major problem, one that I cannot take up in the present essay: the problem of the transformations and the persistence of a lexicon, the historical racism that is in the process of being defined in a new context by the developments in genomics.[8] But for now I would like to emphasize another point: the area of Parel was the cradle of a flourishing textile industry that long formed much of the basis of Mumbai's economy. The crisis of this industry in the 1980s and 1990s, within a process of long contestation by the great workers' struggles, has transformed Parel into a good example of a "wasteland of the dispossessed," with the ruins of factories visible from the windows of Wellspring Hospital. As Rajan (Rajan 2006: 97) brilliantly demonstrates, it is precisely the "dispossessed," once employed in these factories, who form the bulk of the army of "volunteers" on whom Genomed conducts its experiments: "Just as Marx describes the forced proletarianization of the working class in the Industrial Revolution in volume 1 of *Capital*, so one can see how forced deproletarianization as a consequence of the crippling contradictions of capitalism leading to the virtual death of an entire industry in Mumbai leads in Parel to the creation of a new population of subjects who are created as sites of experimental therapeutic intervention."

"Comrades, let us speak of the question of ownership . . ."

Rajan quite rightly insists on the fact that these "experimental subjects" are the subjects on whom the *labor* of the development of this customized medicine is based, a form of medicine that the neoliberal individual aspires to consume. He also draws our attention to the fact that in his research, he found that while the latter tends to hold a US passport, the former tends to have an Indian passport, demonstrating that the elements of continuity with the "old story of colonial expropriation of the resources of Third World" are, at the same time, the elements of a radical discontinuity with that same history (Rajan 2006: 281). To mention only one other major thematic that cannot take up here, but to give at least a sense of its complexity, I will simply quote from Rajan's work: "The relationship of India to the United States as I am trying to configure it, therefore, is not the relationship of an outside to an inside (a binary or relativist framing from which no project of strictly symmetrical comparison can completely escape) but the story of the outside that is always already within the hegemonic inside—but within it in ways that make the inside uncomfortable, distend it, but never turn it 'inside out'" (83).

This is Rajan's singular and effective description of the real and characteristic earthquake of our political, economic, and cultural geography that is currently being exposed by the processes of globalization. But I must begin to conclude, and thus I will go back to the citation by Roberto Esposito from which I began, to his insistence on the relationship between the "inside" and "outside" as the critical point of "biopolitical investigation," as well as his insistence on the need to prioritize the "forces" over the "forms." It seems to me that by reference to the Marxian analysis of primitive accumulation we can productively complicate the picture, opening new horizons for research and indicating at least one of the directions in which to move in order to update the toolbox with which critical thought can orient itself to the analysis of contemporary capitalism.

I mentioned before that the "experimental subjects" of Parel are the subjects on whom *labor* bases the production of "biovalue," the valorization of capital invested in the development of biotechnology and "postgenomic" pharmaceuticals. I should clarify: I do not intend with this formulation to revive some variant of the theory of labor value, as if the exploitation of this labor was in itself sufficient to explain the origin of the huge profits of "biocapital." On the contrary, I am aware of the dead-end immediately encountered by such theoretical proposals. The fact is that biocapital directly exploits—besides labor—the terrain of the *bios* itself, the terrain of a *commons*, the appropriation of which is a key moment in the economic dynamics I am discussing. *Appropriation*, as is well known, *precedes* the juridical institution (the "form" in Esposito's terms) of private property; it is clearly no accident that this institution is subjected to the powerful tensions proper to the terrain of the *bios*—against, for example (but it is indeed only an example!), the development of the so-called biobanks.[9]

Now the relationship between "inside" and "outside," which is problematic to define in terms of "space," seems to assume paradoxically "tempo-ral" characteristics—configuring the *bios* more as an "elsewhen" than as an "elsewhere" of capital. But we can perhaps assert, with the necessary caution, that one of the classic problems at the center of the Marxian analysis of "so-called primitive accumulation"—the violent imposition, through the enclosures, of the norm (once again: the "form") of private property onto the terrains of the *commons* (in the case of Marx, literally the land)—is today one of the fundamental characteristics of the everyday functioning of capitalism. This applies, as we can see, to "biocapital," but also to the land, water, and other traditionally "common" goods.[10] This is especially true for the enormous wealth of knowledge accumulated and produced in networks, to which is connected the powerful problem of the tensions to which this wealth of knowledge is subjected today—once again under the pressure of the emerging *forces* whose power [*potenza*] we must be aware of—the "for-mal" regime of intellectual property.[11]

These are little more than hints, I am aware. But it is my hope that these thoughts are at least sufficient to prompt an agenda for research into a series of issues that seem to me extraordinarily urgent. Others can be added along the same line of reflection, starting with the problem of incomes from rented real estate in the major metropolitan areas and that of the global financial markets, which Christian Marazzi has theorized with his usual effective-ness. Once we emphasize the importance that the commons (in its various forms) assumes in contemporary capitalism's valorization process, we can deal with and map the antagonisms in which these processes are rooted. Thus we begin to take hold of the definition of a theory of the commons different from the simple opposition between public and private property that has dominated modern political and legal thought[12]—a theory of the commons materially incarnated in these antagonisms and able to articulate that "right to wealth itself" that Marazzi has discussed in his work (see for instance Marazzi 2009: 17–49; 2010: 17–60).

From this point of view, the great Foucauldian question of subjection and subjectivation (the *production of subjectivity*) fully recovers its proper central-ity, but once again I think it will be necessary to complicate and enrich this reference to Foucault by taking up a second crucial issue analyzed by Marx in the first volume of *Capital*—the production of *labor power* as a commodity. In a book written some years ago, Paulo Virno (1999: 122–130) pointed out that from the viewpoint of biopolitics, the Marxian category of labor power is absolutely pregnant with meaning. Here I merely want to emphasize how this category signals the radical *scission* that marks the constitution of subjectivity in capitalism, and how it thus presents itself as a critical function in confronting certain readings of contemporary governmentality and biopower—as in the case of Paul Rabinow and Nikolas Rose—which seem to recognize only a *single* model of subjectivity.

In conclusion, like Esposito, I also think that the problems we face require us to grasp "the ways in which the forces are exploding the forms" (2006: 7). But I am convinced that it is equally crucial to understand the multiple ways through which the "forms" are imposed on the "forces" with various levels of violence, capturing them and directing their movement. As I think I have shown, perhaps a bit rapidly but with some plausibility, it is precisely this moment of capture that plays such a decisive role in the functioning of contemporary capitalism. This is a capitalism, let us remember, that spreads out its own heterogenous chains of value on a global scale, continuously remolding the world. Thus we ought to begin to ask (or re-ask) ourselves if the problem is not, in fact, that of the new *creation of the world*. So let us be conscious of the fact that creating the world means "immediately, without delay, reopening each possible struggle for a world, that is, for what must form the contrary of a global injustice against the background of general equivalence. But this means to conduct this struggle precisely in the name of the fact that this *world* is coming out of nothing, that there is nothing before it and it is without models, without principle and without given end, and that it is precisely *what* forms the justice and the meaning of a world" (Nancy 2007: 54–55).

Notes

[Trans.] This article was originally published in Italian as "Le forze e le forme: Governamentalità e bios nel tempo del capitale globale" in *Posse*. Note that the term *forza* in Italian is translated here as "force," but in the Marxian term *forza lavoro* (German: *Arbeitskraft*), the standard English equivalent is "labor *power*." The fact that this differential between *force* and *power*, which occurs in English, does not occur in Italian (in Mezzadra's use of "le forze") should be kept in mind here. The translator would like to thank Andrea Righi, as well as Sandro Mezzadra, for their corrections and suggestions.

1 For a general introduction to the themes of bioeconomy, see Bazzicalupo 2006b and Fumagalli 2007. My usage in this text—in particular in the last two paragraphs—of terms like *biocapital* and *biovalue* is, however, more circumscribed and less ambitious, and specifically refers to the valorization of capital invested in the development of biotechnology and "postgenomic" pharmaceuticals.

2 Two examples are Ferguson 2006 and Ong 2006. On this point I would also like to refer to Mezzadra 2008.

3 For a different reading of this Foucauldian concept, see the essays collected in Chignola 2006.

4 I critically discussed the broad work of Rabinow and Rose in a text written with Brett Neilson, *Border as Method; or, The Multiplication of Labor* (2013). The following considerations stem from this text.

5 But note as a whole, the fourth chapter of this work, "Biopolitics and the Urban Environment." On Foucault, biopolitics, and colonialism, see, among the many texts that could be cited, Stoler 1995.

6 Of fundamental importance for the study of "democracy in India" that I refer to here are the two volumes of Samaddar 2007.

7 I refer to my rereading of chapter 24 of the first volume of *Capital* and published as an appendix in Mezzadra 2008 and to the bibliographical references contained therein.

8 For an initial approach to this problem through Foucault, see the essays of Thomas Lemke (2003), Jürgen Link (2003), and Clemens Pornschlegel (2003), as well as Rose 2007: 155, which seems in this respect, however, too concerned with providing a "reassuring" image of contemporary biopower.

9 For an initial discussion of this problem, see Rodotà 2006, especially chap. 5.

10 Regardless of whether one agrees or not (and I tend to not agree) with the specific project of ecological democracy around which her work is oriented, one should see the analysis of Shiva 2005.

11 For an initial discussion of the problem, see Benkler 2007.

12 Hardt and Negri 2009 is rich in ideas in this sense.

References

Bazzicalupo, Laura. 2006a. "Economia e dispositivi governamentali." *Filosofia politica* 20: 43–55.

Bazzicalupo, Laura. 2006b. *Il governo delle vite: Biopolitica ed economia.* Rome and Bari: Laterza.

Benkler, Yochai. 2007. *The Wealth of Networks: How Social Production Transforms Markets and Freedoms*. New Haven, CT: Yale University Press.

Chatterjee, Partha. 2006. *The Politics of the Governed: Reflections on Popular Politics in Most of the World*. New York: Columbia University Press.

Chignola, Sandro, ed. 2006. *Governare la vita: Un seminario sui corsi di Michel Foucault al Collège de France (1977–1979)*. Verona: ombre corte.

Esposito, Roberto. 2006. "Storia dei concetti e ontologia dell'attualità." *Filosofia politica* 20: 5–12.

Fumagalli, Andrea. 2007. *Bioeconomia e capitalismo cognitivo*. Rome: Carocci.

Ferguson, James. 2006. *Global Shadows: Africa in the Neoliberal World Order*. Durham, NC: Duke University Press.

Hardt, Michael and Antonio Negri. 2009. *Commonwealth*. Harvard University Press.

Legg, Stephen. 2007. *Spaces of Colonialism: Delhi's Urban Governmentalities*. New Delhi: Blackwell.

Lemke, Thomas. 2003. "Rechtssubjekt oder Biomasse? Reflexionen zum Verhältnis von Rassismus und Exklusion." In Stingelin 2003: 160–83.

Link, Jürgen. 2003. "Normativität versus Normalität: Kulturelle Aspekte des guten Gewissens im Streit um die Gentechnik." In Stingelin 2003: 184–205.

Marazzi, Christian. 2009. "La violenza del capitalismo finanziario." In *Crisi dell'economia globale. Mercati finanziari, lotte sociali e nuovi scenari politici*, edited by Andrea Fumagalli and Sandro Mezzadra, 17–49. Verona: ombre corte.

Marazzi, Christian. 2010. "The Violence of Financial Capitalism." In Fumagalli and Mezzadra 2010: 17–60.

Marx, Karl. 1990. *Capital*. Vol. 1. New York: Penguin Classics.

Mbembe, Achille. 2003. "Necropolitics." *Public Culture* 15: 11–40.

Mezzadra, Sandro. 2008. *La condizione postcoloniale: Storia e politica nel presente globale*. Verona: ombre corte.

Mezzadra, Sandro, and Brett Neilson. 2013. *Border as Method; or, The Multiplication of Labor*. Durham, NC: Duke University Press.

Nancy, Jean-Luc. 2007. *The Creation of the World; or, Globalization*, translated by François Raffoul and David Pettigrew. Albany: State University of New York Press.

Ong, Aiwha. 2006. *Neoliberalism as Exception: Mutations in Citizenship and Sovereignty*. Durham, NC: Duke University Press.

Papadopolous, Dimitris, and Niamh Stephenson. 2006. *Analyzing Everyday Experience: Social Research and Political Change*. New York: Palgrave Macmillan.

Pornschlegel, Clemens. 2003. "Die Gegenwart der Eugenik: Zum 'Fall Perruche.'" In Stingelin 2003: 206–27.

Rabinow, Paul, and Nikolas Rose. 2006. "Biopower Today." *BioSocieties* 1: 195–217.

Rajan, Kaushik Sunder. 2006. *Biocapital: The Constitution of Postgenomic Life*. Durham, NC: Duke University Press.

Rodotà, Stefano. 2006. *La vita e le regole: Diritto e non diritto*. Milan: Feltrinelli.

Rose, Nikolas. 2007. *The Politics of Life Itself: Biomedicine, Power, and Subjectivity in the Twenty-First Century*. Princeton, NJ: Princeton University Press.

Samaddar, Ranabir. 2007. *The Materiality of Politics*. 2 vols. London: Anthem Press.

Sanyal, Kalyan. 2007. *Rethinking Capitalist Development: Primitive Accumulation, Governmentality and Post-Colonial Capitalism*. London: Routledge.

Shiva, Vandana. 2005. *Earth Democracy: Justice, Sustainability, and Peace*. Cambridge: South End Press.

Stingelin, Martin, ed. 2003. *Biopolitik und Rassismus*. Frankfurt: Suhrkamp.

Stoler, Ann Laura. 1995. *Race and the Education of Desire: Foucault's History of Sexuality and the Colonial Order of Things*. Durham, NC: Duke University Press.

Virno. Paulo. 1999. *Il ricordo del presente: Saggio sul tempo storico*. Torino: Bollati Boringhieri.

Nietzsche in Contemporary Biopolitics

Tazaki Hideaki
Translated from the Japanese by Gavin Walker

Where Life and Politics Meet

To begin, how are life and politics related to each other? *Bios politikos* indicates precisely the construction of the political through life, through ways of living. But on the other hand, merely living, that is, life as what is common to all living beings, is generally considered to be something originally unrelated to politics.

For example, the *polis* as "public realm" theorized in Hannah Arendt's (1998) *The Human Condition* is, above all, the site of liberation from the limitations and necessities that originate from the fact that the human is a living being, and is thus the space wherein freedom can be realized. This is precisely what Foucault (1978–86) points out by emphasizing in the first volume of the *History of Sexuality* that beginning in the eighteenth century, living itself, the fact of living existence, became the object of politics. For

positions 27:1   DOI 10.1215/10679847-7251871
Copyright 2019 by Duke University Press

Foucault, power until this point—represented by that of the lord—took the form of the power to kill. Then, in the eighteenth century, this became the power to "let live," that is, power began to "care" for the living, a form of power he called "biopower."

However, we should not misunderstand this point: of course, biopower is not unconnected to death—it also kills. Foucault himself attempts to describe the difference between the power to kill and the power to "let live" by suggesting that biopower is "discarded in death." In other words, Foucault's theorization of biopower strongly implies, therefore, that life and politics encounter each other for the first time in the eighteenth century, and that from that point forward, politics subsumes life. It is thought that he argues this with regard to the historical shift from the power to kill to the power to let live, and, certainly, this is the way Foucault's analysis has been taken.

But when politics subsumes life, we must consider the question of what precisely this "life" indicates. Indeed what sort of "thing" is this life? What form does life take at the moment when it is touched by politics? In the "Critique of Violence," Benjamin (1978) calls it "bare life," a life reduced to itself, divested of adornments and form. The object of biopolitics is precisely life in this sense. It goes without saying that today the thinker developing the most profound theorization of "bare life" is Giorgio Agamben. In *Homo Sacer* (1995), he expands on Foucault's concept of biopower, developing a study of biopolitics in line with Benjamin, and above all, with Carl Schmitt.

There are also those who have argued against this way of posing the question of "bare life." For a Deleuzian like Maurizio Lazzarato, for instance, politics and life do not encounter each other somewhere in a certain site, but rather mutually inhere in each other from the very outset. For those who argue against "bare life" in this manner, posing the question in terms of a site in which politics and life meet simply repeats a schema of substance and appearance, that is, perhaps they consider that this posing of the question is not far from saying that life as a still-formless material is simply given form through politics. They might argue that just as life itself has no dynamism without taking form, perhaps they simply bandy about this phrase "bare life" (in fact, Lazzarato is more careful than this; see Lazzarato 1996, 1999, 2000).

Certainly, in Foucault's description too, the inscription of power on the body might prompt us to conceive of an image of power as something sadistic and yet external. For example, in *Bodies that Matter*, what irritates Judith Butler (1993) about Foucault is precisely this sort of inscription as discipline: the image of power operating on the body as if it were hot wax imprinted by the stamp of a seal.

For the Deleuzian, the problem here is Foucault's blurring. What Deleuze expresses in terms of "desiring machines" is the immanence of life, while Foucault frequently counterposes pleasure to desire, that is, he treats pleasure as though it were a raw material that ought to be utilized to its full extent. (In Deleuze's conception, the masochist is a master of the immanent theory of power. In contrast, what should we make of Foucault's obsession with sadomasochism? Although he grasped the *énoncé* itself as an immanence or virtuality, can this endure?)

Throughout these investigations a number of questions are being asked after: what can life do, what can the body do, what potentials do life and the body hold? Is the force that life nurtures external to it or immanent in it? In other words, is biopower, the power to make something live, an immanent potential in life, or is it a force that intervenes from the outside? If we consider that from a certain point in time, a force that was previously external to life has now begun to intervene in it, this becomes a theoretical question concerning the apparatuses of power that are separate from the potentials of the body. Thereby this force called "power" becomes something with entirely different origins from those of life—but it is precisely Nietzsche who insists on something different from this way of grasping the question of power.

This is the context in which we read Nietzsche today.

America/Europe; or, How Is Bare Life Discovered?

In a rather descriptive style, Immanuel Wallerstein's (1974) *The Modern World System* distinguishes between the "world economy" and the "world empire." In the early modern period, the possibility of a "world empire" existed within Spain, but it collapsed and, to be brief, for the first time in the modern period, a "world economy" was established. For Wallerstein, a world empire existed prior to this, but a world economy—economy as world—did not exist prior to the advent of modernity. "World economy" here means the world capitalist system. Wallerstein's critical point is that the capitalist economy was a world capitalism from the very outset of its establishment, and that only on the basis of world capitalism as a premise, can we conceive of something like a "national economy." Thus the political counterpart to the world economy is not a single "empire," but a systematic balance of power among sovereign states.

In order to more exhaustively investigate Wallerstein's argument, we should imbricate and cross-read his discussion with Carl Schmitt's (2003) *The* Nomos *of the Earth in the International Law of the* Jus Publicum Europaeum.

Here, Schmitt theorizes the mode through which the Catholic world imperial order of medieval Europe shifted to an order based on the balance of power between sovereign states (within European territory), in relation to the appropriation of space from the non-European world (in this case, the oceans and the "New World"). Schmitt, who absolutely detested both legal positivism and the theory of natural law, grasps the nomos, the law that cannot be dissolved by a positivistic legality (*thésis*), as the appropriation of space, the appropriation of continents, the division and distribution of land and establishment of place. At the basis of all order is the partitioning and distribution of concrete space. In early modern Europe this problem emerged in the form of the appropriation of the "New World." The civil wars of Europe, which took the form of religious warfare, paralleled the disintegration of the imperial appropriation of space.

Schmitt argues that the greatest contribution of the modern system of sovereign states can be seen in its legitimation of war, that is, the fact that states wage war according to the law places war under legal controls and, in this sense, turns war into something human. At this point, the medieval Christian theory of the holy war was relatively quiet—it was rather through the differentiation of the "correct enemy" from the criminal that war became something other than the punishment of criminals. The "correct enemy" was no longer the target of a campaign for annihilation.

The other side of the process through which domestic Europe was ordered in terms of equal relations between sovereign states, and the civil wars of Europe were concluded, was the wild irruption of violence exerted in the land appropriation of the colonies. For Schmitt, the partitioning of space between Europe and the "New World" signified precisely the differential between a situation with law and a situation without law. The theoretical discussions of the shift from the state of nature to that of society as a result of the social contract began to emerge in the early modern period, but in Hobbes, Locke, and even Rousseau, the image of the "state of nature" at the time tended to be marked by the shadow of the situation of the "New World."

Schmitt praised and lauded the greatness of Hobbes—this is because it was precisely Hobbes who grasped the fundamental nature of the appropriation of land and described the state of nature *as a state of war*. For Locke, the state of nature is not war. As was the case in the "New World," the primordial natural world freely handed over its abundance to man (of course, it was not the indigenous population of the New World, but rather the European colonizers who fully "utilized" this fertility of nature). Thus, Schmitt argues that Locke treats the state of nature as something peaceful precisely because at the time of his analysis, the European appropriation of land had already reached a certain stage of completion. After the principle of the distribution and division of land had already been established, the question of how much the latent abundance of nature can manifest itself has its roots in the fact that it comes to be seen only in the form of the "peaceful" order of competition between the dynamic (colonizer) and the impotent (native), that is, the fact that people are positioned as "equal" competitors is itself an act of violence for the appropriation of land, for the disintegration of society.

Hobbes assumes that the state of nature is a certain type of logical fiction rather than a primal state of man, that is, precisely as what remains when only the social itself is subtracted. It has been widely pointed out that this imagined "destruction of society" has a certain correspondence to the actual disintegration of societies in reality. Thus it has been understood in many cases that in the background of this concept is the civil war in Europe, and in particular, the bourgeois revolution in England. In contrast, Schmitt points out that without the existence of "America," or more accurately, without the existence of a clear dividing line between the couplet "America/Europe," Hobbes's theoretical position is rendered impossible.

But what does this mean? First, the presumed relation between the state of nature and the state of society in the theory of the social contract is, in fact, not a temporal division, such as something historicized in terms of before/after, but rather a spatial division. This is the representation of a spatially constructed difference as something historical, that is, the representation of the state that exists beyond a certain spatial line as a past that one has already experienced. In other words, at the same time as it is a representation of one's own existence, of being oneself, it is also a representation of the other state over there as something one has already overcome.

Schmitt considers that Europe has survived its civil war precisely through this mechanism. The disintegration of the imperial order was overcome through imperialism. In the Western Hemisphere, it is not simply that the invaders, the colonizers, committed genocide against the native peoples or turned them into slaves. Among the invaders themselves there developed a mutual murderousness that was impossible within Europe itself. Through its correspondence to this space of violence, Europe for the first time was able to accomplish its own regional peace. This accomplished, on the one hand, the concentration of power in the sovereign, the sovereign's legitimate monopoly on violence, as well as the delegitimation and criminalization of private violence within the territory of a single sovereign. On the other hand, it installed a legitimation of violence between sovereign entities, that is, the legitimation of war.

However, the significance of "America" for political theory is not limited to this formation of the sovereign state. Rather, we should consider it as follows. In the nomos, that is, the appropriation of space, it is not only that the society of those whose space is taken comes to be destroyed but also that for the appropriators of space themselves, in as much as they also thereby destroy their own society, it is the appropriation of land that renders the establishment of the nomos itself possible (this is what Marx refers to as "primitive accumulation").

If Hobbes were to treat the "New World," America, as the "state of nature," it is not because, as Locke believed, there remained in America an untouched form of nature. The "America" that existed prior to its invasion by Europeans is not the "state of nature." On the contrary, the state of nature in Hobbes indicates precisely the Europeans who invaded "America" themselves.

In Benjamin's (1978) "Critique of Violence," he speaks of a "law-founding violence." Of course, if we casually translate this conception, it will seem simply like the power that establishes the legal system. We do not know to what degree Benjamin was familiar with Schmitt's argument when he wrote this text (in contrast to his later *The Origin of German Tragic Drama* [1998]), but here Benjamin is touching on something closely linked to the discussion Schmitt develops in relation to the nomos, that is, he is dealing with the violence that sustains the nomos. This *Gewalt* must be simply and directly translated as "violence."

Here Benjamin brings up the example of war. He asks, what happens when the areas seized through warfare are recognized as formal territories through cease-fire agreements or peace treaties? The violence of war provokes a transformation in the factual situation in the form of occupation. But this is not all. At the same time, its power forces the factual situation into the law, that is, war forcibly transforms the situation into something normative.

Violence is located at the border between the factual and the normative, or to use a different set of terms, it is located at the border between the constative enunciation and the performative enunciation. That is, it is neither factual nor normative. It is neither descriptive nor performative. It belongs to neither of these two terms, yet nor is it located in the outside. Violence is the degree zero shared by the factual and the normative, the singular point at which the two cannot be differentiated (or what Kierkegaard called the exception, the singular).

In precisely the same way that the deeds of Jesus as charismatic figure are simultaneously the destruction of law and its ultimate fulfillment, that is, just as his deeds were both factual and also of the law, we can see that the law, violence, and life are located in site wherein they cannot easily be distinguished from each other when he states "I came not to send peace, but a sword" (Mt. 10:34).

Of course, Schmitt, who thought that violence or law could be incarnated in the charismatic life of a single living being, and Benjamin, obsessed with the aggregated bodies composed by the workers on a general strike or the viewers in a movie theater, designated this singular point by different names, reflecting their differences in direction: the sovereign (Schmitt) and the oppressed (Benjamin). And yet, it was certainly Schmitt who went the furthest down a path divergent from that which grasps the law neither simply as something normative, nor as something grounded in the factual (for example, the attempt to derive the legitimacy of the law from the common presuppositions of human language ability, as in a "universal pragmatics"). Precisely because of this, the trajectory from Benjamin to Schmitt (Schmitt of course, lived far longer than Benjamin, and was able to see in detail the transformations of law and war in the aftermath of the Second World War) has been deeply emphasized, not only by Agamben but also by thinkers such as Norbert Bolz and Samuel Weber, and has become a point of serious contention within contemporary thought.

This is because, above all, it is a perspective that we cannot overlook in terms of thinking the problems of contemporary biopolitics. If we want to clarify the ways in which the capitalist economy, developing on a world scale (the state of nature), and the agents of its "global nomos"—race, the state, international bodies, NGOs—are becoming intertwined and enmeshed in each other, Schmitt remains a necessary thinker for us.

Now, Hegel understood this contradistinction between the state of nature and the state of society as the contrast between civil society (economy) and the state.[1] In Hegel, the transition from civil society to the state is on the one hand conceived as a type of historical shift, but at the same time, is thought as a transition in constant, daily repetition, that is, as a becoming (the becoming of the state from civil society, the becoming of politics from the economic spirit). In this sense, we can say that the state of nature and the state of society are not merely stages of development, but rather exist in a topological relation to each other (and the condition of possibility for civil society itself is the tearing away of the individual from the "natural" order of the family).

As Leo Strauss (1987) argues at the end of his *The Political Philosophy of Hobbes*, Hegel is not so much the successor to Descartes as he is the successor to Hobbes—he is a thinker of the state of nature that obtains in the wake of Europe's appropriation, division, and ordering of the world. *The Phenomenology of Spirit* (Hegel 1977) begins from the depiction of "life made bare," that is, from "sense certainty" (*sinnliche Gewissheit*). What these deitic terms like *now, here, I,* and so forth indicate is an existence that is no more than "speaking being," something that is nothing but a living thing possessing words, that is, something that possesses nothing other than the fact of being human, the human being as an existence robbed of all else but itself, the most void sense of life possible. *The Phenomenology of Spirit* is a description of the process by which this life made bare once again discovers itself—through the mechanism of reciprocal self-recognition (*gegenseitige Anerkennung*)—within the community called the state. We "moderns" can survive the experience of civil society only by carrying within ourselves a slave, through self-enslavement. For Hegel, the state is above all the community that is created by those who have survived civil society and is conceived as the sole relief measure from the economy as global existence.

Hegel's presupposition is also "America." In "America" the indigenous population is torn apart from their space, captured as a group into the interior of imperialism, their bodies employed as slave labor power, their souls convertible to Christianity. These are the people who barely survive as the economy encircles and seizes the globe—even today, this is not a thing of the past. We need to highlight the appropriation of space that proceeds under the topology of America/Europe, which occurs whenever the IMF, the World Bank, or various governmental bodies attempt to capture people as a set in the form of "developmental potential" in order to think the question: on this earth, is there any space of refuge from the world economy beyond the state?

"Bare life" does not indicate some element that exists nowhere else, an existence simply of a form of life that waits to be molded and processed by a function that arrives from the exterior. Certainly, bare life is something whose expression of correlation, whose site of operation is a biopolitics, a politics that refers to life. But the power of biopolitics is not something *added* to life from the outside, something that processes and transforms it. Rather, it is located in the power of subtraction, so to speak. It is neither something added nor something suppressed. Rather, it is the power to make human life manifest itself as a pure potentiality—precisely what Marx referred to as "labor power."[2]

For Hegel—and probably for today's communitarians—this sort of bare life is considered something that cannot exist without being given expression by the state, or at least by some form of community (besides humanity itself). This is because they think that biopolitics in the final analysis gives something to this life made bare—the power to survive, or the result of having survived? Indeed what Nietzsche incessantly fought against was precisely this sort of understanding of life and power. Of course, it is not the case that Nietzsche's thought opposes the contemporary recognition that our lives are being made bare. Rather, what he would be irritated by is precisely the absence of a politics capable of grasping this increasingly bare life as a moment of potentiality (for example, see "What the Germans Are Missing" in Nietzsche 1997).

Neurosis, Force, Interpretation

From what perspective does Nietzsche problematize the question of life? Or rather, perhaps we must begin with the question of what "perspective" itself is for Nietzsche. Nietzsche (1968: fragment 481) states that facts do not exist, all that exists is interpretation, calling such a standpoint "perspectivalism." Thus, he also rejects the substantialization of the interpreter. What is interpreted—we can perhaps call this the "text"—and what interprets do not form a relation of mutual exteriority. Just as in the strange depiction in Wittgenstein's *Tractatus* (1922), when a text depicts a perspective that belongs to the world, in the world there also belongs a perspective that interprets this text.

Of course, Nietzsche refers to this act of interpretation as desire, drive, or indeed, as will (to power). This can be seen precisely as the constructivist standpoint beloved of so-called "postmodernism." That is, Nietzsche's argument is that the world, or phenomena, can only be thought in correlation to the perspectives (desire, drive, will) through which it is constructed. But is it not the case that this conception of being as making, as production, is, to borrow Heidegger's language, fundamentally determined by Western metaphysics?

It is perhaps rather strange to consider that the utmost form of the Western metaphysics of constructivism is so highly regarded precisely as if it were an overcoming of Western metaphysics. Nietzsche's project for the "inversion of Platonism" has been criticized for being nothing more than an "inverted Platonism." Here we see the criticism that the "overcoming" of Western metaphysics is in fact precisely the "completion" of Western metaphysics, but a doubt remains beyond this: is the path to "overcoming" beyond "completion"?

But did Nietzsche truly conceptualize this schema of overcoming through completion? Certainly we cannot say that it was not there at all. He repeatedly put forward a "plan" for his complete works and then started from scratch again, repeating this act of rewriting. But what should we try to discover in this repetition, in the massive fragmentary manuscripts he left behind? This inability to complete a single work is itself a declaration of divergence from the sense of historicity in which something is completed, and only then another begins.

Nietzsche always remarks that his writings have no readers. He complains that no ears exist to hear his voice. Above all else, Nietzsche's is a voice that states, "This cannot be heard by anyone," a voice that is recorded or documented but that cannot be understood. That is, what characterizes Nietzsche's voice is precisely this noncontemporaneity, not only in the sense that he himself was out of step with the age in which he lived but in the sense that this is a voice that does not conform to the actuality of any age at all.

Earlier, we encountered this perspectivalism in which the interpreting standpoint inheres in the interpreted object. How can we reconcile this inherence of perspective with the noncontemporaneity of the listener?

Here let us turn to the relation between interpretation and the neural. We must escape the spell that binds us to the confirmation of the empirical. The discursive method of constructivism, which argues that being is a production, that truth is (socially) constructed, allows us to entirely elide the question of power relations.

What is crucial in the question of power relations is that this way of posing the problem is something like a hinged metaphor, related simultaneously to the factual confirmation in the sense of physical relations of force, as well as the sense of the performance of relations of order and submission. For Nietzsche, interpretation must be grasped in terms of power relations. In other words, meaning is not something that is constructed as some kind of object through a single perspective, but rather perspective becomes something inevitable.

For example, when we consider the symptoms of an illness, we take into consideration a variety of interpretations: interpretations on the nuclear level, interpretations of molecules, the interpretation of certain bodily tissues, the interpretation that is performed by the body itself, the interpretation of one's own consciousness, the interpretation by a physician, and so forth. One molecule, one worm, one painting, one enunciation, one piece of poetry, one human being. All of these things are each interpretations of the universe as a whole, or the interpretations of conjoined phenomena (which are themselves interpretations).

For Nietzsche, relationships of cause and effect too, are nothing more than interpretations. That is, what appears to be factually-empirically confirmed is in fact something performative, something that follows the sequence of order and submission. Thus, the entirety of the universe can be described in terms of relations of power in this sense of the relation of order/submission. Of course, it is not quite so simple. Our gesture of reinterpreting this word *power* from a metaphor into something literal and one-dimensional, was itself already interpreted by Nietzsche in advance. Moreover, this "in advance" itself is untraceable.

We, the readers of Nietzsche, who say "this too is an interpretation," cannot be the contemporaries of this Nietzschean voice that is inscribed in the text in advance—we are too late for this voice. Through our mechanically repeating voices that state "this too is an interpretation," a statement that has lost all meaning (because it cannot be interpreted), our interpretation is overthrown each time we say this, and, as a result, the "interpretation of interpretation" that is posed in the question "what is interpretation?" becomes impossible.

In this situation in which we find ourselves, a situation in which we cannot interpret whether or not power is a factual relation or a performative one, in which we cannot interpret the question "what is interpretation?," we might as well call this sense of interpretation by the name "life"—it is placed at the degree zero between the factual and the normative.

Beginning with "neurosis," the "quantitative relations of force" and so forth, the terminology of "natural science" belovedly employed by Nietzsche indicates precisely this site of the degree zero, the image of this impossibility known as the "interpretation of interpretation." The (stimulation of) neurosis is another expression for a Dionysian intoxication, that is, it itself is a certain interpretation of this relation of the Dionysian and the Apollonian. In other words, it is an interpretation of visual representation (to which the metaphor of "perspective" also belongs) and the aural (voice). Neural stimulation is a phenomenon that belongs to an entirely different order from that of visual representation or aurality—something that cannot be configured as forming a causal relationship between the two—yet it is the process of translation of these mutual elements. The neural is a translation between phenomena, the interpretation of interpretation itself. But precisely because it is something entirely natural to us, we cannot see or hear it—that is, it is not a phenomenon, but being (of course, it is not the Other World or *Hinterwelt*). And further—although I have repeated this many times—it is neither factually verifiable nor merely performative.

Toward a Neuro-Biopolitics

Grasping life (made bare) as a pure nervous system—understanding the being (of the world) as a set of neural links: a neuro-bio-ontology. It was not only Nietzsche but also his contemporary (roughly two years his senior) Daniel Paul Schreber, who in fact *lived* this ontology (see Schreber 2000; Santner 1996). Does there exist a politics of the neurally wounded, the neurally suffering?

Today, the (welfare) state promises to protect our nerves, to heal them. "Your nerves are sensitive, they can't handle too much excess stimulation, so don't worry, the state will protect you from all excess." "You mustn't create any neural links with anyone else, you never know what sort of excess stimulation might result. So I, the state, will mediate all your neural links, serving as a form of resistance, a barrier intervening between them." We see what we want to see, and what we do not want to see is excised from our nervous system, its stimulations shut out. The welfare state, constructed by the media and its technocrats, formulates apparatuses of protection against our nervous systems, increasingly being made bare on a world scale by the processes of globalization.

If "testimony" today is descending into crisis, it is precisely because we are increasingly covered over, enclosed, and simultaneously made bare by these apparatuses. Testimony is above all a question of the nerves; it is nothing more than the "interpretation of interpretation"—the transformation of that which has been seen into words. Testimony is the communication of the impossibility of the interpretation of interpretation (for example, what is communicated to us when we read Nietzsche's literary legacy). But we cannot endure this neural stimulation so we block it out—we suppress this stimulation, treating representation only as representation.

Nietzsche's literary legacy is a sort of survival—not merely a survival after the death of Nietzsche himself, but also in the sense of having barely survived the experience of Nazi ideology—and lending our ears to this testimonial, the survival of this fragmented voice, is a form of training or practice for us toward the creation of a set of neural linkages entirely different from those delimited by the state's capacity for representation (pathos, suffering). Indeed, it is precisely the emergence of a neural politics from within the biopolitical that furnishes the meaning of our own linkage to Nietzsche today.

Notes

1 [Trans.] In Japanese, the term "civil society" (*shimin shakai*) retains the dual sense of the German term *bürgerliche Gesellschaft*, which in English can be rendered both "civil society" and "bourgeois society." This terminological point should be kept in mind in the discussion of "civil society."

2 On the question of potentiality, see Agamben 's "On Potentiality" (1999: 177–85).

References

Agamben, Giorgio. 1995. *Homo sacer: Il potere sovrano e la nuda vita*. Turin: Einaudi.

Agamben, Giorgio. 1999. "On Potentiality." In *Potentialities: Collected Essays in Philosophy*, edited and translated by Daniel Heller-Roazen, 177–85. Stanford, CA: Stanford University Press.

Arendt, Hannah. 1998. *The Human Condition*. Chicago: Chicago University Press.

Benjamin, Walter. 1978. "Critique of Violence." In *Reflections: Essays, Aphorisms, Autobiographical Writings*, translated by Edmund Jephcott, 277–300. New York: Schocken Books.

Benjamin, Walter. 1998. The Origin of German Tragic Drama. London: Verso.

Butler, Judith. 1993. *Bodies that Matter: On the Discursive Limits of "Sex."* London: Routledge.

Foucault, Michel. 1978–86. *History of Sexuality.* 3 vols. Translated by Robert Hurley. New York: Random House.

Hegel, G. W. F. 1977. *The Phenomenology of Spirit*, translated by A. V. Miller. Oxford: Oxford University Press.

Lazzarato, Maurizio. 1996. *Videofilosofia: La percezione del tempo nel postfordismo.* Rome: Manifestolibri.

Lazzarato, Maurizio. 1999. "Gabriel Tarde: Un vitalisme politique." In vol. 1 of *Œuvres de Gabriel Tarde: Monadologie et sociologie*, 103–50. Paris: Institut Synthélabo.

Lazzarato, Maurizio. 2000. "Du biopouvoir à la biopolitique." *Multitudes* 1, no. 1 : 45–57.

Nietzsche, Friedrich. 1968. *The Will to Power*, translated by Walter Kaufmann and R. J. Hollingdale. New York: Vintage.

Nietzsche, Friedrich. 1980. *Sämtliche Werke: Kritische Studienausgabe*, edited by Giorgio Colli and Mazzino Montinari. 15 vols. Berlin: de Gruyter.

Nietzsche, Friedrich. 1997. "What the Germans Are Missing." In *Twilight of the Idols*, translated by Richard Holt. 43–49. Indianapolis: Hackett.

Santner, Eric L. 1996. *My Own Private Germany: Daniel Paul Schreber's Secret History of Modernity.* Princeton, NJ: Princeton University Press.

Schmitt, Carl. 2003. *The* Nomos *of the Earth in the International Law of the* Jus Publicum Europæum, translated by G. L. Ulmen. New York: Telos Press.

Schreber, Daniel Paul. 2000. *Memoirs of My Nervous Illness.* New York: New York Review of Books.

Strauss, Leo. 1987. *The Political Philosophy of Hobbes.* Chicago: University of Chicago Press.

Wallerstein, Immanuel. 1974. *The Modern World System.* Berkeley: University of California Press.

Wittgenstein, Ludwig. [1922] 1981. *Tractatus Logico-Philosophicus.* London: Routledge and Kegan Paul.

The Financialization of Knowledge and the Pooling of Populations: On the Biopolitical Management of Geocultural Areas during the Transition from Industrial to Bioinformatic Capitalism

Jon Solomon

Introduction

The crisis in the abstractions of risk calculation that erupted in 2008 in the US housing market, spreading like wildfire across the global financial markets, was rooted in what Brian Holmes (2010) calls the "artificial world model" of financial capitalism that led to a "fundamental disconnect" between information and event. For the postcolonial critic, Holmes's diagnosis immediately calls to mind a much older phenomenon common to colonial history. Whether in imperial Japan's invasion of China (Tanaka 1995) or Britain's governance of Palestine and the Middle East (Said 1979), there have always been crucial moments in the history of modern colonialism when the institutional production of authoritative knowledge about the colonial social formation had become so disconnected from the actual process of social composition that it actively contributed to policies ending in violence

positions 27:1   DOI 10.1215/10679847-7251884
Copyright 2019 by Duke University Press

and/or disaster. Naturally, such examples of "disconnect" are not limited to colonial governance but are part of the modern problem of knowledge about the social and governance in general. What is so interesting—and dangerous—about the present time is that the various forms of "disconnect" appear to be merging in a single matrix. Rather than indulge in outmoded fantasies of decisive epistemological vigor that would once and for all resolve the dreaded problem of "disconnect," this essay aims to chart out a genealogy of the problem and describe a fundamental link between the "artificial world model" of financial capitalism and that of what are called in North America "area studies." Although it may take some time before the financialization of knowledge in process will induce the same kind of crisis as that seen in 2008, due to the financialization of capital (after all, the former process occurs at a slower pace relative to the latter), the "artificial world models" propagated by each have become as inextricably intertwined as they are disconnected from the growing indemnity of destruction that can never be repaid. Hence, even as we strive to imagine long-term solutions, we must also devise strategies for surviving the violence of the geocultural area.

The Archaeology of Modern Areas

As I have described in previous work (Solomon 2010), Foucault's *The Order of Things* (1966) presents the aporetic relation between knowledge and experience as one of the defining "methodological horrors" of the modern epoch. The root of this methodological aporia derives from the amphibological status of "man." "Man" is "a being such that knowledge will be attained in him of what renders all knowledge possible" (Foucault 2002: 347). It is not the attempt to make "man" into an object of science, to objectivize human existence, that is for Foucault the fundamental defining trait of the modern archaeological strata. It is rather the attempt to pair that effort with a parallel attempt to experience the limit of knowledge per se that defines the modern figure of the human. Between these two efforts lies a constant tension between experience and knowledge that defines the anthropological situation of modernity. The attempt to reveal "the conditions of knowledge on the basis of the empirical contents given in it" (Foucault 2002: 347) leads to a series of unavoidable yet irresolvable oscillations that the various different systems of modern thought, from positivism and empiricism to dialectical negativity and phenomenology, each try, with only relative success, to master. These would be, in short order, the oscillation between the empirical contents of experience and the transcendental limits to knowledge; the oscillation between an analysis of the positivist type ("the truth of the object determines the truth of the discourse that describes its formation" (349) and a discourse of the eschatological type ("the truth of the philosophical discourse constitutes the [promise of the] truth in formation" [349]); the oscillation between a nature (that determines experience) and a history (that determines knowledge); and finally the oscillation between an individual body and a collective culture. In the face of this series of fundamental oscillations, modern thought—excluding that alternative undercurrent to modern philosophy emblematized by Spinoza's "savage antinomies" (Negri 1991)—is always looking for a mediating third term. Foucault proposes that this third term always takes the form of "actual experience." The resort to actual experience, however, does not resolve the ambivalence or amphiboly that existed from the outset. "It is doing no more . . . than fulfilling with greater care the hasty demands laid down when the attempt was made to make the empirical, in man, stand in for the transcendental" (Foucault 2002: 350). From a philosophical point of view, the upshot of this amphiboly is a constantly receding, infinite series of displacements in the "miniscule" yet infinite difference that "resides in the 'and' of retreat *and* return, of thought *and* the unthought, of the empirical *and* the transcendental, of what belongs to the order of positivity *and* what belongs to the order of foundations" (370). The more modern thought resorts to the category of actual experience to mediate the weight of the oppositions it has gathered under the aegis of a contradictory singular point—the empirico-transcendental doublet that is "man"—the more it must make an effort to repair the irreparable schism that constitutes "man" from the very beginning. This accelerating circularity, a "dog-chase-tail" syndrome, constitutes the main paradigm—and problem—of modernity as identified by the archaeological method.

It is my hypothesis that the geocultural area characteristic of modernity has been the principal apparatus (*dispositif*) by means of which the methodological horrors of the circular relation between experience and knowledge that constitute the modern-construction of the human have been managed in the modern era. To put it in Foucaultian terms, I am suggesting an archaeological explanation for the apparatus of geocultural areas. The ratio that governs the distribution of the heterogeneity between experience and knowledge—the much sought-after "magic bullet" of modern social theory in general—is what is commonly summoned under the name of "rationality." The irrational is the name given to those forms of experience that cannot be appropriated into normalized, that is, rationalized, forms of knowledge. Significantly, the dominant image of rationality in modernity is projected on an entity that combines both hierarchies of knowledge and taxonomies of social organization in a single geocultural unit denoted, quite improbably, by a directional term, "the West." The apparatus of geocultural region literally attempts to "ground" the oscillation induced by the figure of man. This structural fact may help explain the reason that Enlightenment thought, even when it aims for subjective autonomy through "universal" knowledge, is always an anthropological project, based on hylomorphic ontology, whose goal is to realize perfection of the species in the gendered, ethnicized and racialized terms of a *national* or *civilizational* figure of the human.

But the apparatus of area was not just a solution to a knotty metaphysical problem. The "oscillation" everywhere in evidence today, as Peter Hitchcock (1999: 10) astutely observed a decade ago, is not just the symptom of a theoretical postmodernism but concerns the status of the body and materiality versus knowledge and cognition in the transitions unleashed by capitalist modernity. By releasing the indeterminate flexibility and radical interconnectedness of human labor beyond the classical measure of time while simultaneously intensifying the external means of measure through superexploitation, the contemporary form of cognitive, bioinformatic capitalism is, in this sense, nothing but the complete subsumption into capitalist valorization of the oscillation inherent in the modern construction of what Foucault called "Man." Subsumption means both cultivation and subordination, socialization and proletarianization, subjectivity and subjection. As such, this subsumption would have been impossible without an apparatus of capture that, as Sandro Mezzadra (2011: 304) writes, "establishes *frontiers* in the spaces it invests." This apparatus of capture was most forcefully staked out in the era of industrial capitalism by the creation of "the West" as the template of a geocultural area. It combines *rationality*, a technology for codifying the oscillation between experience and knowledge, with *translation*, a technology for managing the oscillation between different regimes of accumulation and valorization. Today, "the West" continues to be effective precisely to the extent that its supposed universality has been teleologically particularized. While it is doubtful that the social formations we conventionally call the West can continue to maintain their dominance, their downfall in itself would not mean that the amphibological structure of the geocultural area initiated by the birth of "the West" (which was *not* a "Western" phenomenon) as an apparatus of biopower will cease to reproduce itself.

At the end of an essay devoted to a reflection on the problem of the apparatus, Giorgio Agamben (2009: 24) concludes that the machine of governmentality, which delivers us unto a proliferation of apparatuses, is "leading us to catastrophe." Since I have elsewhere discussed the relation between catastrophe and the particularly modern apparatus that is the geocultural region (Solomon 2009b), let me briefly summarize: the geocultural area is intrinsically connected to catastrophe inasmuch as it is intended to be both a form of response to the catastrophe that has already taken place (deterritorialization or the *terra nullius* effect unleashed by colonialism) and a measure of prevention against possible future disasters (reterritorialization in the form of the nation-state and civilizational difference). Yet by its very nature the apparatus of the geocultural area prolongs and deepens the crisis of unequal development and destruction. As Couze Venn (2009: 221) reminds us, unequal development does not happen because of a lack of action (the lack, for instance, of proper development policies) but is rather the result of an active process, "a process of producing the poor" through expropriation. The apparatus of the geocultural area (and the subjects it generates) is one important part of this actively productive process. For this reason, we must admit that the catastrophe of which Agamben speaks is not of the order of the possible but rather of the actual. It has already taken place—starting with the formation of "the West" as the model of a leading global region that combines or sutures different translations of value in order to produce an anthropological figure of rationality.

Now, we have to account for the way in which this apparatus of area is produced at the same time that it is productive. The first step in my response to this challenge is to recast our understanding of who or what is the subject of area studies. Returning once again to Agamben's discussion of the Foucaultian concept of apparatus, we place great importance on the crucial role of "the subject." The subject is that which mediates between "living beings" and "apparatuses" (Agamben 2009: 14). In the case of the apparatus called "area," no social identity after the "native" epitomizes the subject more than the *specialist*. My methodological assumption, following the work of Gilbert Simondon, is that relation as a process or operation, called "transduction" by Simondon, mutually coproduces each of the separate terms that appear to precede, and hence constitute, the relation (Combes 2013; Scott 2014). Hence, in order to think the subject of the apparatus of area, we need to understand how a process of individuation in the preindividual common of bodies, tongues, and minds simultaneously produces *both* natives *and* specialists. We need, in other words, to understand epistemological *specialization* and social *speciation* from the perspective of the relationship between the two.

The link between nativity and competence begins with the commodification of labor. Competence, as both cause and effect of the division of labor, legitimizes that division. Nativity, as both cause and effect of the organizational form of the nation-state, legitimizes the system within which that divisional structure occurs. The challenge for critique lies in being able to show how the two are mutually generated through a singular process or operation. One possible way of thinking this operation is suggested by the growing consensus around primitive accumulation not as an historical "stage," but as an account of how capitalism organizes and distributes the social narrative of its incessant transitions (Read 2003; Mezzadra 2011; De Angelis 2007; Walker 2011). The key problem here is that the transition itself produces material effects on the constitution of knowledge (a major element in the formation of subjectivity), and that knowledge, concretized in material practices, institutions, languages, and discourses, further produces an effect analogous to that of a cause with regard to the transitional event.

In the case of the apparatus of area, this abstract formula of mutual feedback between causes and effects means that we need to account for a generalized vision of racism that goes beyond "race" to account for the "milieu" that sustains it. To be more precise, we need to be on guard against every determination of both the anthropological essence of the "human" and the various conclusions drawn from that determination that are institutionalized in the distribution of anthropological difference in parallel fashion across both the disciplines of the humanities and the political organization of human populations. When Foucault speaks of racism in a biopolitical context, he is not referring to prejudice against shared characteristics or traits, but rather to the normalization of heterogeneous elements in a single discursive field. If we think of normalization as a decisional strategy of individuation, rather than a measure of inherent value, it becomes possible to see, somewhat counterintuitively, how the specialist becomes the subject of an apparatus called geocultural area. This means, of course, that we do not just want to look for the rules that govern normalization in the practices of everyday life in a specific region, that is, we cannot simply just "study the natives" and their putative racial traits and/or prejudices in any particular area, but we must also pay attention to the mode or operation by which an area is mapped onto a cognitive field. In other words, we need to account for the *relational mode* that creates *both* the specialist *and* the native (including, of course, admixtures of the two such as the native specialist or the nonnative nonspecialist specialist of other disciplines) at exactly the same moment. This change in perspective means positing that the specialist is the necessary subject without which the ideological naturalization of modern areas could never be possible, precisely the figure that activates indifference to the oscillation between experience and knowledge and enables naturalizing attempts to ground modernity through the categories of anthropological difference such as the native. Yet it is also important to recall that the creation of indifference is, according to Jason Read's (2003: 108) analysis of Marx's account of capitalist transitions, also part of the very condition of the initial abstractions—the wage, abstract labor, and the commodity—that constitute, "the capitalist mode of production extending itself to other regions of the world." Indifference is a subjective technology of capitalist abstraction designed to manage the phenomena of bordering and transitioning essential to capitalist regimes of accumulation. To say it another way, the production of subjectivity is the very site at which material processes of bordering, extension, and transitioning in the commodification of labor are managed. Because the geocultural area is not just a way to mediate the unwieldy division between the empirical and the transcendental at the heart of "man," but also forms a technology for displacing the ceaseless borderings and transitions of capitalist "development" onto the battleground of subjectivity, the geocultural area provides a necessary complement or "habitat" for the modern construction that is "man." The specialist, an operator of that matrix, is, as much as the native, an anthropological figure par excellence, the crystallization of a whole series of presuppositions about what constitutes the essence of the "human" and how to legitimize population management in a (post)colonial context based on that figure. But above all, the specialist is a tool to leverage the transitions in the mode of production into the mode of subjection.

The main function of the apparatus of area thus lies in the legitimation of *social speciation* through which subjects can identify themselves as autonomous with regard to the mode of production, that is, "the production of ways of life and subjectivity necessary for the production of value" (Read 2003: 111). Yet that "freedom" to be autonomous comes at the price of a new necessity. As long as the "decision" to constitute the apparatus of geocultural area can be related to necessity, be it through birth, competence, or obedience (141), it is impossible to imagine the individuations of bodies, tongues, and minds and their assemblages of a milieu in a way other than the geography of anthropological difference that characterizes the regimes of accumulation spawned by colonial-imperial capitalist modernity. Rather than approach specialization and speciation as two separate moments, the one pertaining to a division of labor based on the specificity of the cognitive, the other pertaining to a question of anthropological difference based on the specificity of organic embodiment, we need to find the crux before the two moments are distributed and reproduced—both cognitively and materially. We need, in other words, to bracket the *point of decision*,[1] the place where the division between saying and doing, cognition and materiality, is posited (only to be rejoined later). The focus here is on grasping the relation between epistemological specialization and social speciation without assuming either as norm.[2] What follows in this essay is an attempt to uncover not only the contemporary genealogy of that point of decision but also the ways in which its apparent necessity has been retroactively superimposed and is today undergoing further radical transformation.

With this in mind, let us now examine how the restructuring of the university that has been in process since the beginning of the 1990s has dramatically affected the *milieu* in which area studies are practiced. The key to understanding this dramatic change is to be found in the articulation of financialization, bioinformatic technologies, and population management.

The Financialization of Knowledge: Institutional Logics

Although the *commodification of knowledge* under the regime of industrial capitalism was quite well established, the new conditions of cognitive capitalism are creating a situation that differs significantly, bearing important consequences for the biopolitical management of "area" and "population" required by the modern state. What we have today, in addition to the commodification of knowledge, is a new situation that we might call *the financialization of knowledge*.

Generally speaking, financialization is a technique of recentering the actions of many around future value speculation at a time when the traditional measure of value—labor time—has been disrupted by cognitive forms of production (Negri 2013) or displaced by the valorization of debt (Lazzarato 2012). The key component of financialization that I would like to highlight concerns the assignation of value in view of future risk within a context in which value has intrinsically become highly volatile. The process of financialization can be assigned to two logics, one institutional, the other informational.

As concerns the institutional aspect, a brief inventory may suffice. The *financialization of the university* refers to the whole series of adjustments to labor conditions (Gulli 2009), capital investment (Ross 2009), student debt (Bousquet 2008), and property rights (Rossiter 2007) within the university that have changed the definition of fundamental players: knowledge has been transformed from a national good into products with marketable value; students have been changed from a future source of labor into a direct source of undervalued labor; faculty have been transformed from extensions of the social elite, the arbiters of national aesthetics, to precarious, exploited labor or the representatives of corporate interest; research has been transformed from expanding disciplinary horizons to the production of patents and accumulation of excellence indicators. These changes, which began in the Anglophone universities and spread quickly to the East Asian universities and elsewhere around the globe, are belatedly occurring in Europe under the banner of the Bologna Process, begun in 2001, and the Doha round of negotiations over liberalization in trade and services under the WTO. In the eyes of Andrew Ross (2009), the financialization of the university is leading to the total transformation of the university into a "knowledge corporation" consolidated amid the redefinition of higher education, under the WTO, as a global service industry.

The Anglophone university, pursuing a global market agenda that has essentially usurped debate about the social value of education within the WTO framework, is the institutional locus in which the two most salient contemporary trends in the humanities and social sciences have developed: the first being the multiplication of interdisciplinary structures through a plethora of topical "studies" programs; the second being the wholesale elimination of departments and programs amid a general restructuring of the university according to parameters established (and irregularly flouted) by the corporate-oriented evaluation bureaucracy. Amid such vast changes in the organizational and disciplinary structure of the contemporary university, the role of the humanities deserves to be questioned for its relation to the current historical transition in the mode of capital accumulation. My hypothesis is that contemporary interest in interdisciplinary studies per se does not constitute a challenge to the apparatus of area, but is rather a way of managing adjustments in the nature and function of areas as devices of labor control and capture during a period of historical transition in the regime of accumulation.

Like all manner of "studies" that have mushroomed over the past several decades, interdisciplinary studies has left the essential disciplinary structure of the humanities unchallenged. Here I am not talking simply about the divisions between sociology, literature, and history, but rather the differences among different types of subjectivity ascribed to the producers of knowledge and the mapping of that difference according to geocultural areas and the schema of anthropological difference. If the colonial division within humanity described by Osamu Nishitani (2006) between *anthropos*, a provider of raw data, and *humanitas*, a historical subject of theory devoted to the constant renovation and progress of knowledge, is defined by the stance one adopts toward knowledge, then it follows that the production of subjectivity in the colonial-imperial modernity has to be understood not simply as the product of communal practices that can be the object of ethnographic study, but as always already implicated in the production of knowledge in general. Within this scheme, every act of knowledge production eventually relates in the final instance to the global index of anthropological difference. The "performativity of indexing" (Sakai 2010: 446) names that aspect of humanistic knowledge when it is seen as a social practice within the context of the new forms of population management associated with capitalist development and global colonization. Seen from this perspective, disciplines look different. The objects, methodologies, and theses that constitute each of the modern disciplines are themselves always already "anthropological commentaries" on the supposed essence of human being and community that implicitly "position" the subject within processes of social segmentation related to capitalism and global colonization. Interdisciplinary studies, rather than proposing a radical break with this structure, inscribe themselves into what Brett Neilson and Sandro Mezzadra (2013) describe as the *multiplication* of labor associated with flexible border regimes, and in that sense is probably the one form of activity in the humanities that is more symptomatically critical of the apparatus of area today than any other. But the question has always been whether multiplicity displaces control.

In a recent work devoted to the philosophy of posthumanism and critical animal studies, Cary Wolfe (2010) astutely observes that in the institutional deployment of interdisciplinarity, "border crossing" is never undertaken by the disciplines themselves, but is rather assumed to be the business of individuals. This keen observation can be fruitfully related to Max Haiven's (2014: n.p.) acute assessment of the way in which the neoliberal university localizes its problems in the figure of the individual through a process of *externalization*, "the downloading of a systemic and structural crisis onto the lonely, precarious individual." Decades of personal observation leaves me with no doubt that the costs of supporting transcultural intellectual production are overwhelmingly borne by individual researchers and their partners and families, while the labor of intellectual workers devoted to transcultural studies proceeds largely along the model of R&D outsourcing to consumer-producers characteristic of post-Fordism: "the amateur just loves to work, with no concern for compensation or ownership of that work" (Kleiner 2010: 17). For this reason, the individual—or rather, a certain form of individuation favored and sustained by the apparatus of area—is undoubtedly the site where the corporate restructuring of the university and the institutions of interdisciplinarity converge.

Caught between economic *externalization* and ontological *inflation*, the individual is a key linchpin in the system of interdisciplinarity that ideologically masks the restructuring of labor and knowledge in the age of global biosemiotic capitalism. Needless to say, this "individual" itself is a fabrication whose ideological dimensions parallel those of the objects that mediate social relations and serve as the fetish of disciplinary knowledge. The process of individuation corresponds to the fetishization of disciplinary objects, culminating in the consolidation and marketing of the university-based knowledge worker's personal brand and the assignation and accumulation of value through the tools of financialization that dominate higher education today. Meanwhile, the disciplines are not left simply to lumber along as before, but, under pressure from the new forms of measurement of academic labor that aim to standardize and discipline the extraction of value, they must adopt a posture of *retrenchment*—a synonym for *return*—for increasingly frantic academic productivity hidden under the rhetoric of "platforms" and "collaboration."

My hypothesis is that traditional disciplinary studies, such as area studies and sociology, and interdisciplinary studies *considered simultaneously together*, constitute one of the ways in which *the industry of the means of production* has been externalized, foisted onto the individual, who must bear its costs. The commodification of human communication, not just of this or that particular language, but of the communicative faculty of the species in general, is an oft-noted characteristic of the post-Fordist era (Virno 2004; Agamben [1996] 2000). When we take into account Paolo Virno's description of how the "communication industry" today plays a role that was traditionally given over to the industry that produced machines and other material necessary for production in general, we begin to see that area studies and interdisciplinary studies actually play a similar key role, in spite of their complete marginality, to the operation of today's transnational capitalism. This does not mean that area studies and interdisciplinary studies are intrinsically part of capital: the reduction of academic labor to measures of value recognizable by information technology is not required to enable forms of communication that produce transcultural social value while refusing the process of capitalist valorization and its ideological reflection in the culturalist schema of cofiguration, but *is* required to control and profit from it. What is intrinsic to post-Fordist capitalism is, as Virno (2004: 54–66) has argued, a certain appropriation of the performativity of species-being in general. Transculturalism itself is not an end product (in fact, it should be, strictly speaking, indistinguishable from the act of producing transculturally), yet the university system imposes from the exterior of transcultural production the demand for end products that can be reduced to numerical values. For Virno, the capture of species performativity for the extraction of surplus value defines the post-Fordist moment.

Institutions of National Translation and the Indexing of Performativity

Today's corporate university is devoted to the *indexing of performativity*. New forms of measurement have been grafted onto academic labor according to a post-Fordist model of flexible labor management guided by burgeoning evaluation bureaucracies. Performance is constantly being measured, evaluated and finally indexed in a series of ranking systems oriented toward competition in a global market. Increasingly large amounts of time devoted to supervision and evaluation are required of intellectual laborers. Yet the notion of a *global index* is, as we have seen, nothing new. Its roots lie in the modern discourses of anthropological difference emerging out of the synergy between colonialism, capitalism, and the new sciences of biology, philology, and political economy. Anibal Quijano (2000) and Naoki Sakai (2010) have both emphasized the way in which colonialism calls forth a mobile measure of relative social status on a global scale through the categories of anthropological difference. For Quijano, race performs the role of a global classificatory system that can be applied in any specific local context to produce a quick and clear reading of relative superiority and inferiority. This "reading" is essential to a mystification of the commodification of labor. Hence, the global index of anthropological difference has, since the beginning of European colonialism, always been related to the process of valorization (the social effects of which are seen most visibly in the division of labor but would also include the valorization of exchange value as the sole, exclusive value). The missing piece in Quijano's elaboration is the way in which the racial indexing of labor populations necessary to global capitalism has relied on the production of subjective resistance to capture value. Sakai's work on translation, subjectivity, and the disciplinary divisions of knowledge draws attention to that previously neglected yet crucial aspect. I would like to be allowed here to insist on the importance of Sakai's work in allowing research to finally identify the relation between the process of valorization and the index of anthropological difference through *two parallel operations of translation*. The first is the translation of use value and social value into exchange value; the second is the translation of social difference into specific difference—what is today commonly called "cultural difference," but must certainly also be understood to include "gender difference" and "sexual difference" as part of a larger configuration of anthropological difference in general. Together, these two forms of translation constitute the *apparatus of area*.

The capture of surplus value—the commodification of labor—simply could not proceed without an analytical center that stabilizes and regularizes the production of subjectivity through the translation of social difference into anthropological difference. The modern university has traditionally played the role of being precisely that sort of authentication center. The compilation of premodern civilizational archives, museums, and monuments and their codification into discrete disciplines whose mission was to define the aesthetic, affective component of national class difference is precisely the historical situation described by Herbert Marcuse ([1964] 1991) in *One-Dimensional Man*. In this sense, universities were the command centers of *national translation*—the production of knowledge into the national idiom according to the limits of the apparatus of area and the granting of social recognition only to the subjects attached to that apparatus. The democratization of the university in the twentieth century did not fundamentally alter this paradigm, as it was still based on a Fordist and essentially national mode of production. The global index of anthropological difference associated with the organization of labor through the apparatus of area remained unchallenged. What is new today, in relation to the global index function, are the plethora of technologies that radically transform the temporality and spatiality of indexing practices. Needless to say, these innovations have threatened radically to undermine the territorially based capture of value. Henceforth, capitalism, now in a "Great Transformation" from an industrial to a cognitive model, has had to create entirely new devices to capture value.

This change has posed great difficulties for the Institutions of National Translation (INT) such as the university. Before exploring these difficulties, I would like to explain what the term *INT* refers to. Basically, the INT is a way of interpreting what Bill Readings (1996) has famously described as "the University of Culture"—one of the two great models of the modern university, the other being the techno science model. In my reading, the University of Culture is a national institution charged with the task of "translating" all experience—especially the experience of anthropologized temporality that covers the heterogeneity of the past, the experience of populations in the present that are supposed to serve as the nation's "others," and the return to national essence teleologically projected onto the future—into knowledge, and all knowledge into and out of nationalized idioms, while at the same time legitimizing in a general way the domestic (i.e., national) division of labor at the basis of social class. Beyond the crucial role the university plays in the creation of national culture and civilizational difference (not to mention its role in the division of labor and gender inequality), its other purpose is theoretical or rational: to institutionalize and to regulate the ratio that constitutes the paradigmatic quasi-object of modern spatiality—those complex mixtures of experience plus knowledge that we know as geocultural areas. The INT has thus been fundamentally linked to three types of translation essential to the creation of modern regions: the creation of national culture as a reference point for human "social speciation" in a field of hylomorphically constructed anthropological difference; the establishment of a reasonable ratio between experience and knowledge that would grant relative stability to the taxonomy of the social; and, finally, the translation of social value into surplus value, seen primarily in the reproduction of social class.

Today, under the impetus of the great transformation currently under-

way, the possibilities for relying on the aesthetic, or affective structure, established by the nation-state through the technology of national language to
naturalize humanistic knowledge production within a supposedly homogeneous cultural situation (for which the university is as much a cause as
an effect), are being vastly reduced. For this reason among others, the post-
Fordist version of the *indexing of performativity* has deeply exacerbated the
fundamental (post)colonial anxiety reflected in the *performativity of indexing*.
At the same time, it has also created a situation in which transculturalism is
not thought—much less organized and funded—on the basis of the inde-
terminacy of language(s) and people(s)—but is rather formulated according
to the logic of specific difference. Colonialism meets post-Fordism exactly at
the point of systemic complicity between the area studies/interdisciplinary
studies binome and transnational, neoliberal, bioinformatic capitalism. Due
to the postmodern/postcolonial disappearance of stable national standards
that would permit a unique set of values to be transmitted or translated, as
was the case under conditions of industrial capitalism, into knowledge commodities that can be evaluated (by peer review) and exchanged in a national
market, the value of living goods within the contemporary informational
university, that is, the value of the innovative assemblages of bodies, tongues,
and minds recognized as knowledgeable, is given to wild oscillations that
require new forms of evaluation and new regimes of translation.

The Financialization of Knowledge: Informational Logics

The two primary vehicles of evaluation in the University of Finance are
digitalization and property rights. These forms converge, in the case of the
humanities, in the utilization of commercial indices (e.g. Social Science
Citation Index, SSCI) and normative audit bureaucracies culminating in
global ranking tables and "excellence" initiatives that favor social policies
of "differential inclusion" (Mezzadra and Neilson 2013). In spite of the fact
that the commercial indices are currently heavily weighted toward Anglophone journals, it is clear that the long-term trend is to include other languages such as the Asian national ones through a variety of avenues such as
the establishment of parallel rating systems (such as TSSCI in Taiwan), the
initiation of local language editions of SSCI-listed journals, and the admit-

tance of local journals. Much like the ratings offered by credit agencies, audit bureaucracies in the humanities have become a tool for imposing criteria external to production that would create the collective opinion necessary for the valorization of goods, such as knowledge about various geocultural areas, whose value would otherwise be subject to impressive oscillation in a globalizing era. I am suggesting, in other words, a parallel between the tools of financialization—such as the Black-Scholes formula developed in 1973 to determine the price of derivatives—and the procedures of the audit culture that is becoming the foundation of applying a measure of value on the production of knowledge in the post-Fordist informational university.

One of the main effects of the audit bureaucracy is to externalize the problems associated with the value and the meaning of research in the humanities, not to mention its potential to bring about disciplinary innovation. At the level of the audit bureaucracy, knowledge is not directly evaluated but rather becomes the object of quantification according to specific indicators. What matters is not the level of innovation—which would be all but impossible to measure quantitatively—but rather the number of publications, citations, awards, patents, and so on, that constitute numeric values in league table rankings. Value is derived from the relations and links among the digitalized work, its circulation, and a general connection to a network of other works equally deemed "valuable." We might say that sociality itself constitutes the site of valorization. In the process of digitalization, however, the quantitative indicators recognized by the audit bureaucracy no longer carry any meaning beyond their value as indicators.

As Matteo Pasquinelli (2015: 3) points out, we are dealing with a "topological shift" that is concerned not so much with the "open flows of information" as the "accumulation of *information about information*, that is metadata." The evacuation of meaning, or its translation into metadata, effectuated by the evaluation system plays an important role in the context of globalization. The point where this becomes really interesting for social theory is to be found in the nexus between population and area that culminates in the apparatus of the geocultural area. The significance of this development, however, appears to be lost on a field of knowledge still very much indebted to the anthropological legacy inherited from the colonial-imperial modernity.

Informed by the framework of anthropological difference inherited from colonialism, these "areas" have largely been structured in the postcolonial era by debates over the meaning and ethics of so-called cultural difference. The political stakes of these debates are directly related to the global division of labor and the value of migrant labor in general. What these debates invariably obscure, however, is *both* the way in which so-called cultural difference per se was historically constructed as a hylomorphic model of anthropological difference/normative model of governance for the world system of nation-states that emerged out of imperialism *and* the way in which *the logic of so-called cultural difference now serves an entirely different purpose in managing, like financialization, the wild oscillations that characterize the chaotic social transitions from industrial to cognitive capitalism.* Hence, Read (2003: 151) observes: "communication in real subsumption plays the role of the wage in formal subsumption—it is the simultaneous site of mystification and struggle." The provisional conclusion that Read draws from this situation is to confirm the mobilization of "a particular bad infinity, in which this contradiction is repeated again and again in different theoretical languages and registers" (151). The oscillations inherent in capitalist valorization meet the oscillations inherent in the amphibological construction of modern "man" and its corresponding "habitat"—the apparatus of geocultural area. Contemporary bioinformatic capitalism aims for nothing less than the full mobilization of this situation.

What we are seeing today is not simply the collapse of measure that was first described by the Italian autonomists but also a reintensification of its means. Under post-Fordism, there is in fact an oscillation at the heart of the process of valorization, an oscillation that moves roughly between the poles of "immaterial" and "immiserated" labor (Dyer-Witheford 2002). Cognitive capitalism needs a way of simultaneously cultivating and disciplining the oscillation between the two. It needs, in other words, an ideology. So-called cultural difference supplies the necessary index, but what is so interesting is that it is not just an index but also a form of organization (for both production and consumption) and value intrinsic to post-Fordist production itself. Here is where the greatest extension of labor time (what Marx called "absolute surplus value") is performed: the cultural codes that are necessary for production are extended to every facet of life, constituting one of the primary means through which a cognitive measure can be imposed on post-Fordist production.

To that end, something like a transnational cognitive bureaucracy, with its own peculiar forms of etiquette (Solomon 2014), is in the midst of emerging. If, as Yann Moulier Boutang writes, "the encounter between the most developed communism-of-capital on the planet [i.e., the United States] and a communism-reduced-to-a-jacobinism-of-the-bolshevik-bureaucracy [i.e., the People's Republic of China; my translation] is no surprise," what continues to be a cause for surprise is the way the two ostensibly separate social formations that are developing, in spite of (or really because of) the discourse of cultural difference that would seem to separate them, a unique cognitive bureaucracy devoted to managing binomial constructions: the privatization of "Chinese" bodies, tongues, and minds, on the one hand, and the pollution of "global English" ones, on the other. This division corresponds to the difference between "immiserated" and "immaterial" labor much more than nationalized ethnicity; beyond the relative waxing and waning of US imperial nationalism, global English has already become internal to the reproduction of class difference within "Chinese" social formations. The primary task of the transnational cognitive bureaucracy is, as studies of network culture have alerted us, *not* to create a single hegemonic meaning (such as would have been the goal of the INT under conditions of industrial capitalism) but rather *"to increase the effectiveness of the channel"* (Terranova 2004: 14) and maximize signal-to-noise ratio. The basic formula is the normalization of all cultural phenomenon according to the logics of intellectual property and informatization. This perspective will help us understand the kind of complicity that occurs in the midst of competition between the United States and the People's Republic of China, as the governments of each state race against each other to find ways of instituting "back doors" on all communications technology that would enable mass surveillance.

Just as Naomi Klein's (2008) investigations revealed a decade ago, growing market links and parallel conceptual modeling between the state security apparatuses of the United States and the People's Republic of China, the consolidation of the transnational cognitive bureaucracy is best investigated as a result of unacknowledged complicities in the realm of cultural production. It is here that the complicity between the burgeoning, state-supported discourse on "cultural security" in the People's Republic of China configures itself in relation to elaborate discourses on "cultural difference" and disciplinary divisions in the corporate university. The "global university," an emerging global market in educational "services," combines two registers—the cognitive and the material—into a single apparatus of population management. The key tools of leverage here are intellectual property rights (IPR) and flexible labor regimes. As IPR becomes the new standard by which measure can be imposed on production, the INT of industrial capitalism has been transformed into a new entity that we might facetiously dub an iPrint® (i.e., an INT that has been requisitioned by cognitive capitalism and mobilized under the guidelines of IPR). My focus here is on understanding a new kind of geocultural distribution. Needless to say, the establishment of INTs occurred considerably later in the states emerging out of colonization than it did in the old imperial centers. Significantly, the slow process of consolidation of INTs in the decolonized nations of the South, often under conditions of prolonged political instability and economic expropriation, has matured just at the moment of capitalism's revolutionary transformation into a cognitive and biospheric dimension. The Academic Ranking of World Universities (ARWU), undertaken by Shanghai Jiao Tong University (SJTU) since 2003, provides an excellent illustration of this historical transformation. Among the four authoritative global ranking systems current today, ARWU is the only one located outside the West; yet it should come as no surprise that the world's lone non-Western ranking system actually reinforces the centrality of status-quo indicators. ARWU places double emphasis on publications in indexed journals compared to that found in the other three (*Times Higher Education*'s World University Rankings, Topuniversities' QS World University Rankings, and *US News & World Report*'s Best Global Universities Ranking). In the first place, ARWU has to rely on Thomson-Reuters for the collection of data indicators (Thomson being one of the main providers of commercial indices); second, the ARWU ranking system project at SJTU was undertaken in close cooperation with the World Bank, suggesting the importance of education to global finance (according to Ross, the global level of trade in services is roughly equal between finance and education). The involvement of Thomson-Reuters, which is both a provider of real-time information about financial markets and a major player in the university-based audit bureaucracy, emblematizes the imbrication of finance and knowledge.[3] As the humanities are restructured into the function of "cultural sampling" (Holmes 2002), imparting a set of linguistico-cultural skills considered necessary to the creative industries in a global environment, the humanistic part of the university preserves its former importance as an INT, but the operation of translation that was at the core of the old INT is displaced from a national subject of citizens' rights to an informationalized subject of intellectual property rights. This retooling is at the basis of the financialization of knowledge under the iPrint® and constitutes a crucial vector in the governmental technologies of population management for what Jacques Bidet (2006) calls the emerging "global State."

The Financialization of Knowledge: Logics of Privatization

Transformations in the status of language may help to illustrate the impact of the iPrint® on the apparatus of geocultural regions. While many people have commented on the deleterious effects of English language dominance—what Cronin (2003: 128) unforgettably calls, in relation to the politics of translation, "clonialism,"—scant attention if any has been paid to the way it may affect the reorganization of the public/private dichotomy in relation to property. The association of English with "international" and other languages with "local" contexts imposes a new way of delineating "public" from "private" space along a bifurcation into "global" and "local" access (even while the meanings of "global" and "local" are themselves undergoing profound transformation), conferring an implicitly "privatized" status on other languages associated with the "local." Ironically, we are now witness to an era in which entire nationalized languages—which were themselves historically constructed through the appropriation of exchange in order to legitimate the norms of the class structure crystallized in the state—are becoming progressively "privatized" relative to the "public" use of global English. The recent case in which the Republic of Korea successfully applied to have the Dragon Boat Festival, common in China and other parts of East Asia, recognized as part of its exclusive national heritage under the UNESCO Convention for the Safeguarding of the Intangible Cultural Heritage (2003) demonstrates the extent to which national culture is becoming a form of private property, subject to IPR. Of course, the nation-state since its inception was designed to codify the terms of ownership of the "resources" of particular geographical territories, yet it is fascinating to see how the regime of immaterial labor typical of post-Fordist production is pushing such proprietary claims beyond the conventional, nineteenth-century definition of resources (human, animal, and natural) to extend now to various forms of immaterial property, including, of course, knowledge, training, and, as I would like to emphasize here, language, as well as forms of biological property, such as computer-designed cells and transgenetic manipulations. Given the propensity of neoliberal governments to operate as stakeholders for corporate interests of a complex, (trans)national nature, eroding the category of the citizen, the collective claims implicit in such immaterial resources stand to follow a similar logic. I am not trying to argue against the necessity of a response to expropriation (seen, for instance, in the dispossession of indigenous medicinal knowledge by multinational pharmaceutical companies), but rather for a break with the normalization of bodies, tongues, and minds under any form.

The wholesale privatization of language under the influence of global English is occurring at the very moment when post-Fordist incorporation of a "linguistic model" into the process of production means that language itself is becoming a source of value and proprietary rights. Open access has often been held out as a remedy to the enclosures created by IPR. Yet it is necessary to understand how the question of open access specifically in relation to humanistic knowledge raises issues related to the use of IPR for biopolitical control of bodies, tongues, and minds. Informatization is increasingly blurring the line between the digital and the biological, the organic and the inorganic. In the scientific realm, the overwhelming bulk of open-source material is almost exclusively in English, which merely aggravates a situation particular to post-Fordist production: the "pollination effect" essential to the productive process is virtually monopolized by English, resulting in the concomitant privatization of entire languages.[4] From there it is not very hard to start "pooling" populations for privatizations in other directions, such as gene pools (more about that in a moment). Those advocates of open access whose positions rely on the positive externality of English need to be made aware of what is being excluded: the debate is irrelevant from the perspective of populations/classes outside of English whose access is blocked by biopolitical devices such as the control over tongues and bodies (and genes and prosthetic memory devices such as computers, etc.). The lesson of externalities and immaterial labor is that the only question that really concerns capital is how to figure out a way of enforcing a standard of measure on intellectual products especially when the value of such products—similar to derivatives—fluctuates wildly in a postcolonial/postmodern environment in which it would appear that there is no longer a unique, totalizing system of value guaranteed by the state. Yet appearances are misleading. The new totalizing system is, in fact, premised on the fluctuation of difference. If there is such massive potential value in appropriating or pooling the pollination effect in a single language, it must be understood as the sign of something other than just the same old modern desire for a universal language. What must be emphasized is the fact that English language dominance leaves other assemblages of bodies, tongues, and minds easily exposed to relative proletarianization and privatization, particularly by states whose ruling elites are exactly the ones denying positions to new innovative labor and thus creating the conditions for labor to turn to fundamentalist or nationalist movements. For these reasons, I would think the question of open access, specifically in relation to humanistic knowledge, needs to be displaced toward translingual and transcorporeal pollination against the privatization and proletarianization of bodies, tongues, and minds.

The progressive privatization of national languages "cofigured," as Naoki Sakai (1997) would say, against the domination of global English, is just the tip of the iceberg when it comes to the ways in which externalities, now extending to include virtually all aspects of "life," such as the linguistic and biological being of the human species, are being subjected to total subsumption within the regime of financialization. As an implicit model of translation (Solomon 2009a), global English maximizes the effects of pollination available at any given time. Precisely for this reason, the demand for the imposition of property rights on creative work is felt strongest among the English-language media and publishing industry (where the notion of property is more and more extraneous to the process of production), particularly with regard to the dissemination of information, that is, translation rights, into non-English media and contexts (even as the possibility of "free access" as a business model for English publications appears more and more likely). The biopolitical implications for population management are enormous: just as it can be difficult to identify the population to which a language optimized for pollination such as English corresponds, the correspondence between other, nationalized languages and specific populations is not only substantialized beyond doubt, it also becomes the basis for a limited privatization that extends beyond language to the formation of various "pools" available for capitalist mobilization: the pool of knowledge about a specific population to which a state-sanctioned language/culture supposedly corresponds; the labor/consumption pool; but also equally critically for bioinformatic capitalism, the biotech/pharmaceutical subject test pool, the gene pool, the risk pool, the data pool, and so on.

Pooling: The Real Subsumption of Population

It is a matter of no small importance to find a way to describe the nexus between different tools related to the identification of informational populations—such as algorithms, derivatives, and genetic research—and the logic of cultural difference institutionalized in disciplinary division—such as "area studies" and their respective regimes of translation—that are being used to (mis)manage the unwieldy chaos of the contemporary transitional period.

The phenomenon of "pooledness" begins at the intersection among codes, species, potentials, and accidents. The neat thing about the term *pool* is not only that it spans a wide range of different domains but that it also indicates a fundamental ambivalence that is the site of capitalism's most vexing, and profitable, contradiction. The pool is another name, in the age of bioinformatic capitalism, for *both* the concentration of production and consumption in a continuous feedback loop optimized for value extraction *and* for all the various manner of collective organizational forms, epitomized by the network or by transgenesis (the transfer of genes across species by means of bacterial transport), that do not strictly adhere to the logic of species difference. Pools are not just places where information, resources, and risks are collected and distributed, they are also places where innovation, mutation, and unexpected transport occur. As such, the concept of the pool is emblematic of the central problem of capitalism: how to unleash the creative productive forces through new forms of social cooperation while at the same time disciplining and controlling the forms of cooperation so that they serve the needs of accumulation rather than social, or planetary, needs.

A major characteristic of the logistically pooled population is thus the institution of feedback loops for the goal of expropriation, creating "needs" as much as defining "nature." In a riveting account of the blood trade and the Chinese bioeconomy in the context of global financialization, Anne Anagnost (2011: 233) relates the dangers of one specific biopolitical loop between production and consumption favored today: "bioproducts extracted from bodies may in turn confront those bodies in an alienated and possibly harmful form." Combine this insight with an analysis of the way in which bio-risks are externalized onto the individual and the way in which the pharmaceutical industry and the GMO biotech agro-business purveys products designed to assist populations exposed to potential catastrophe that are themselves derivative of the privatization of life forms, codes, or knowledge sampled from those very same populations, and we begin to have a notion of the contemporary loop that goes way beyond what the 1973 classic film *Soylent Green* (in which a totalitarian state biopolitically manages population crisis through an autophagic alimentary feedback loop: people eating people) had imagined.

The classical definition of *right*, understood as the faculty to exercise choice over a good or service, and the disciplinary apparatuses of quarantining, cannot help us understand the contemporary phenomenon of "pooling" populations that are highly mobile, highly diverse, and also highly monitored. Maintaining bodies, tongues, and minds all coded as "Chinese," for instance, in a state of "pooledness" would be impossible by definition for a single state—even a high-security one—to orchestrate singlehandedly given the gamut of social differences (linguistic, regional, class, ethnic, mobility, etc.) that traverse such a group from the inside/outside; the composition or "pooling" effect is further complicated since it runs across the rift or difficult point of interface between the emergent form of capitalism that is cognitive, or bioinformatic, and the previous form of capitalism that is industrial—a rift that not only runs the international divide but that within each nation as well. While China studies performs the role of witness to the immense multiplicity of Chineseness, it not only reproduces the unity of "Chineseness" (a schema of one-and-many), it also serves to hide the encounter between flows of labor freed by capitalist deterritorialization and flows of capital freed by the commodification of labor through which the anthropological matrix is born.

In truth we are dealing with the phenomenon described by Melinda Cooper (2008) as the neoliberal abandonment of the nation-state model of population control. I understand Cooper's argument not to be that the nation-state ceases to be an important theater of operations but rather that the way in which population as an object of power and a site of surplus value extraction is imagined by the current regime of accumulation is no longer simply based on the metonymic link between theories of immunity and those of sovereignty. To our narrative about the historical transformation in the forms of political organization, particularly the changes in sovereignty and the rising tide of corporate privatization, needs to be added a parallel index of changes in economic production as we move from the industrial to the cognitive, bioinformatic configuration crystallizing today. As Cooper (2008: 19) persuasively argues, this transformation aims to "relocate economic production at the genetic, microbial, and cellular level, so that life becomes, literally, annexed within capitalist processes of accumulation." The creation of an intrinsic link between value production and life, understood as code, has profound ramifications for the organization of the humanities, given that so much of that organization is ultimately indebted to very powerful presuppositions not only about species difference (such as the difference between human and animal) but also about the way in which species difference is related to or reflected in social difference.

What Cooper describes as the historical displacement of surplus value from the commodification of labor to the commodification of code across the boundaries between both the organic and the inorganic as well as those between species might be fruitfully related to the historical transition described by Stefano Harney (2010) as the transition from "statistical populations" to "logistical populations." A notion of disciplinary construction was central to Foucault's understanding of the term *population*. *Population* names an object that is created through the methodological perspective of statistics. The modern nation-state served as the classical locus for population in the colonial-imperial modernity: both the source of surplus value extraction through labor and the legitimation of the state apparatus through anthropological notions of givenness. Logistics, which is concerned with the negotiation of code across barriers, shifts the locus of population from the nation-state to the pool. Its goal is not to avoid the accident (of boundary transgression) but, as in bacterial transgenesis, to figure out how to profit from it. Harney identifies logistics with Marx's concept of "absolute surplus value." What is being extended, is, I would argue, not, as it was for Marx, the length of working hours, but the very division between species, between code as life and code as value. Relative surplus value, which is accomplished by intensive rather than extensive means, comes back into the picture when we start looking at the way in which algorithms and other digital operations essentially create new forms of social aggregation.

If the object that is population appears to have undergone a transformation today, as Harney suggests, then it should be possible to "reverse engineer" the change in order to understand the corresponding restructuring on the disciplinary side of things that correlates to the appearance of a new object. This is the reason that *I speak about area studies in terms of a transnational cognitive bureaucracy that uses techniques of externalization to exploit new arenas of relative and absolute surplus value.* While "area studies" remains tenaciously tied to the geographical index of anthropological difference inherited from the colonial-imperial modernity, it is actually in the business, like the rest of the humanities, of contributing to, as Pasquinelli warns, the construction of an entirely new topography. This topography is generally not visible to the "specialists" who engage in area studies work, nor to the "native social species" they study. It is accessible only from the perspective of the facilities and institutions that handle the "metadata" produced out of academic evaluation bureaucracies. Such topography, based on the algorithmic abstractions of financialization, is capable of mapping movements in everything from gene pools and student migration to recombinant DNA and literary studies, and then further mapping all manner of correlations among the various domains. Hence, while the area studies apparatus was a form of institutionalized nation-state racism under the INT, in today's iPrint® it is becoming part of the bioinformatic circuits of anthropological sampling and autoconsumption within the corporate surveillance state. The remarks made by William Binney, one of the highest-level whistleblowers to emerge from the National Security Agency (NSA), that "the ultimate goal of the NSA is total population control" (Loewenstein 2014), contain an important clue about how the institutional area studies operates today. Just as yesterday's "natives" are being turned into today's pools where the feedback loop between production and consumption can be carefully "optimized" and kept under constant surveillance, so the "specialists" are being transformed into elaborate "face-detection" mechanisms.

I am aware that such claims likely will not sit well with those "specialists" who understand their work through a culturalist ethics of self-and-other or with those "natives" who dream of realizing autonomy and recognition for their "state-ly" otherness, nor will the claims sit well with those "foreigners" who would like to arrogate for themselves the position of conferring, or withholding, recognition for such autonomy. Yet I do not think that in the face of the new topography of algorithmic finance, any ethics or autonomy could be achieved without using one's engagement in the humanistic production of knowledge to also address, plan for, and to militate for, the complete restructuring of the humanities in a way that would be different from the one currently undertaken with the interests of financial capital in mind. The goal is to return what is common to the common, to destroy, in other words, the apparatus of area. Along the way to envisioning this goal, I take a clue from a very interesting proposal, with regard to the division of labor, that comes from the *Multitudes* collective, who advocate the necessity of breaking the link between work and wage as a process of social composition (Lazzarato 2002). Parallel to the radical reorganization of the division of labor, a similarly deep-roots reorganization of the humanities requires that the link between anthropological difference and the disciplinary division of intellectual labor be deactivated in order to highlight the ways in which intellectual production is always already a process of social composition. Pooling, no less than quarantining before it, is not an invention of capital, but rather belongs to the side of living labor. Yet its meaning is fundamentally ambiguous. At once pure subjection and subjectivity as collective power, the rendition of population into "pools" that flow back and forth across the lines between the organic and the inorganic represents capital's hope for a perfect, autopoietic feedback loop in which the entire cycle of life and death will be instantly valorized under the regime of accumulation. Bioinformatic capitalism appears to promise a profitable way of harnessing oscillation, but clearly the parabolic scope of the oscillations is a trajectory of catastrophe. Although many are tempted to look for refuge against these oscillations by returning to the apparatus of area, our investigation reveals that the "solution" itself is part of the problem.

Notes

1   This notion of the point of decision draws on the work of François Laruelle, who has a unique way of approaching what he calls "the philosophical Decision." Ray Brassier provides a helpfully succinct explanation: "All philosophising, Laruelle insists, begins with a Decision, with a division traced between an empirical (but not necessarily perceptual) datum and an *a priori* (but not necessarily rational) faktum, both of which are *posited as given* in and through a synthetic unity wherein empirical and *a priori*, datum and faktum, are conjoined. Thus, the philosopher posits a structure of articulation which is simultaneously epistemological and ontological, one which immediately binds and distinguishes a given empirical datum, whether it be perceptual, phenomenological, linguistic, social, or historical; and an *a priori* intelligible faktum through which that datum is given: e.g., Sensibility, Subjectivity, Language, Society, or History. . . . In other words, every philosophical Decision is a species of what Foucault called the 'empirico-transcendental doublet,' and as such remains a viciously circular structure that already presupposes itself in whatever phenomenon or set of phenomena it is supposed to explain." I take Laruelle's description of the philosophical Decision to be a generalized structure of response to the antinomial oscillation of the anthropological figure "Man" described by Foucault. While the antinomial oscillation itself is not exclusive to modernity, what is specific about the modern response is both the concentration of that oscillation—which has now been shown to us as the essential condition of the neotenous species—in a totalizing anthropological figure that aims to close down the oscillation and, as we have argued here, the complementary attempt to ground that figure in a "habitat" that itself is an amphibological mixture of transcendental and empirical elements, that is, the geocultural area. Laruelle's "nonstandard philosophy" presents us with a series of axioms designed to demobilize being (by preventing any possible convertibility with meaning) and thus protect us from its (de)faults. As such, it is a prophylactic measure that protects us from the oscillation that Laruelle claims is the default position of philosophy in general, while concomitantly liberating a space in which to examine the operation of transduction described by Gilbert Simondon (as seen, for instance, in our attempt to present the specialist and the native as a singular, codependent production).

2 To illustrate what this approach might bring to the table, a task for which I will not have time to elaborate here, it might be useful to distinguish it from the historically important attempts critically to map out power/knowledge configurations institutionalized through area studies under the form of Orientalism. While I do not wish to disqualify the critique of Orientalism outright, since its inception it has always assumed the putative unity of the West as one of its principal avenues of attack. By contrast, once the transductive approach has been adopted, it could be analogously "applied" to any number of different contexts in the humanities. For area studies, probably the most pertinent one will come first with regard to the West/Rest divide that structures debates over the ethics of so-called cultural difference. From this perspective, a series of questions present themselves: Is it possible to account for cultural history in a way that does not assume the preexistence of the entities that constitute the cultural, or translational, encounter? Is it possible, in other words, to see the encounter as that which gives rise to such entities? What will global cultural history look like when we can no longer narrate it in terms of the encounter between the West and its Others? What kinds of adjustments would be necessary in the disciplinary organization of the humanities to accompany and encourage such change? What kind of social and political organization would suit a world that can no longer be mapped and indexed according to the geography of anthropological difference?

3 I would like to take this opportunity to acknowledge my deep gratitude to Paolo Do for sharing his unpublished research concerning the restructuring of higher education in China and its significance for understanding the financialization of knowledge and labor on a global scale. My understanding of the SJTU ranking system, and the effect of ranking tables in general in relation to both labor and financial markets has benefitted tremendously from Do's pathbreaking research.

4 "Pollination" is the name given by economists to describe the mode of production characteristic of post-Fordist immaterial labor when the model of the network increases productivity and innovation through cooperation.

References

Agamben, Giorgio. [1996] 2000. "*Marginal Notes of Commentaries on the Society of the Spectacle*." In Agamben *Means without End*, translated by Vincenzo Binetti and Cesare Casarino, 73–90. Minneapolis: University of Minnesota Press.

Agamben, Giorgio. 2009. *What Is An Apparatus? and Other Essays*, translated by David Kishik and Stefan Pedatella. Stanford, CA: Stanford University Press.

Anagnost, Anne. 2011. "Strange Circulations." In *Beyond Biopolitics: Essays on the Governance of Life and Death*, edited by Patricia Ticineto Cloug and Craig Willse, 213–37. Durham, NC: Duke University Press.

Bidet, Jacques. 2006. "The Rule of Imperialism and the Global-State in Gestation," translated by Jon Solomon. In *Translation, Biopolitics, Colonial Difference*. Vol. 4 of *TRACES: A Multilingual Series of Cultural Theory and Translation*, edited by Naoki Sakai and Jon Solomon, 173–210. Hong Kong: Hong Kong University Press.

Bousquet, Marc. 2008. *How the University Works: Higher Education and the Low Wage Nation*. New York: New York University Press.

Brassier, Ray. 2001. *Alien Theory: The Decline of Materialism in the Name of Matter*. PhD diss., University of Warwick.

Combes, Muriel. 2013. *Gilbert Simondon and the Philosophy of the Transindividual*, translated by Thomas Lamarre. Cambridge: MIT Press.

Cooper, Melinda. 2008. *Life as Surplus: Biotechnology and Capitalism in the Neoliberal Era*. Seattle: The University of Washington Press.

Cronin, Michael. 2003. *Translation and Globalization*. New York: Routledge.

De Angelis, Massimo. 2007. *The Beginning of History: Value Struggles and Global Capital*. Ann Arbor, MI: Pluto Press.

Dyer-Witheford, Nick. 2002. "Composition de classe de l'industrie des jeux vidéo et sur ordinateur." *Multitudes* 3, no. 10: 53–64. English version published as "The Class Composition of the Computer and Video Game Industry." www.multitudes.samizdat.net/Cognitive-Capital-Contested.

Foucault, Michel. 2002. *The Order of Things*. New York: Routledge.

Gulli, Bruno. 2009. "Knowledge Production and the Superexploitation of Contingent Academic Labor." *Workplace*, no. 16: 1–30.

Haiven, Max. 2014. "The Ivory Cage and the Ghosts of Academe: Labor and Struggle in the Edu-Factory." *Truthout*, April 30. www.truth-out.org/news/item/23391-the-ivory-cage-and-the-ghosts-of-academe-labor-and-struggle-in-the-edu-factory?tmpl=component&print=1.

Harney, Stefano. 2010. "The Real Knowledge Transfer." *SocialText Online*, August 30. www.socialtextjournal.org/periscope_article/the_real_knowledge_transfer/.

Hitchcock, Peter. 1999. *Oscillate Wildly: Space, Body, and Spirit of Millenial Materialism*. Minneapolis: University of Minnesota.

Holmes, Brian. 2002. "The Flexible Personality: For a New Cultural Critique." *transversal*, January. eipcp.net/transversal/1106/holmes/en/base_edit.

Holmes, Brian. 2010. "Is It Written in the Stars? Global Finance, Precarious Destinies." *ephemera* 10, nos. 3–4: 222–33.

Klein, Naomi. 2008. "China's All-Seeing Eye." www.naomiklein.org/articles/2008/05/chinas -all-seeing-eye (accessed on December 12, 2010). Originally published *Rolling Stone*, May 14.

Kleiner, Dymitri. 2010. *The Telekommunist Manifesto*. Amsterdam: Network Notebooks. Freely available for download at www.networkcultures.org/networknotebooks.

Lazzarato, Maurizio. 2002. "Garantir le revenu: Une politique pour les multitudes. [To Guarantee Income: A Politics for the Multitudes]" *Multitudes* 8. www.multitudes.net/Garantir -le-revenu-une-politique/ (accessed on May 15, 2014).

Lazzarato, Maurizio. 2012. *The Making of the Indebted Man. An Essay on the Neoliberal Condition*, translated by Joshua David Jordan. Los Angeles: Semiotext(e).

Loewenstein, Antony. 2014. "The Goal of the NSA Is Total Population Control." *The Guardian*, July 11. www.theguardian.com/commentisfree/2014/jul/11/the-ultimate-goal-of -the-nsa-is-total-population-control.

Marcuse, Herbert. [1964] 1991. *One-Dimensional Man: Studies in the Ideology of Advanced Industrial Society*. 2nd ed. Boston, MA: Beacon Press.

Mezzadra, Sandro. 2011. "The Topicality of Prehistory: A New Reading of Marx's Analysis of 'So-Called Primitive Accumulation.'" *Rethinking Marxism* 23, no. 3: 302–21.

Mezzadra, Sandro, and Brett Neilson. 2013. *Border as Method; or, The Multiplication of Labor*. Durham: Duke University Press.

Moulier Boutang, Yann. 2007. *Le capitalisme cognitif: La nouvelle Grande Transformation* [*Cognitive Capitalism: The New Great Transformation*]. Paris: Editions Amsterdam.

Negri, Antonio. 1991. *The Savage Anomaly: The Power of Spinoza's Metaphysics and Politics*, translated by Michael Hardt. Minneapolis: University of Minneosta.

Negri, Antonio. 2013. *Time for Revolution*, translated by Matteo Mandarini. London: Bloomsbury.

Nishitani, Osamu. 2006. "Anthropos and Humanitas: Two Western Concepts of 'Human Being,'" translated by Trent Maxey. In *Translation, Biopolitics, Colonial Difference*, edited by Naoki Sakai and Jon Solomon. Vol. 4 of *TRACES: A Multilingual Series of Cultural Theory and Translation*, 259–74. Hong Kong: Hong Kong University Press.

Pasquinelli, Matteo. 2015. "Anomaly Detection: The Mathematization of the Abnormal in the Metadata Society." Presentation at the Transmediale Festival, Berlin, January 29.

Quijano, Anibal. 2000. "Coloniality of Power, Eurocentrism, and Latin America." *Nepantla*: *Views from South* 1, no. 3: 533–80.

Read, Jason. 2003. *The Micro-politics of Capital. Marx and the Prehistory of the Present*. Albany: State University of New York Press.

Readings, Bill. 1996. *The University in Ruins*. Cambridge: Harvard University Press.

Ross, Andrew. 2009. "The Rise of the Global University." In *Toward a Global Autonomous University: Cognitive Labor, the Production of Knowledge, and Exodus from the Education Factory*, edited by the Edu-Factory Collective. New York: Semiotext(e).

Rossiter, Ned. 2007. *Organized Networks: Media Theory, Creative Labour, New Institutions*. Rotterdam: NAi.

Said, Edward. 1979. *Orientalism*. New York: Vintage.

Sakai, Naoki. 1997. *Translation and Subjectivity*. Minneapolis: University of Minnesota.

Sakai, Naoki. 2010. "Theory and Asian Humanity: On the Question of *Humanitas* and *Anthropos*." *Postcolonial Studies* 13, no. 4: 441–64.

Scott, David. 2014. *Gilbert Simondon's Psychic and Collective Individuation: A Critical Introduction and Guide*. Edinburgh: Edinburgh University Press.

Solomon, Jon. 2009a. "The Proactive Echo: Ernst Cassirer's '*Myth of the State*' and the Biopolitics of Global English." *Translation Studies* 2, no. 1: 52–70.

Solomon, Jon. 2009b. "Rethinking the Meaning of Regions: Translation and Catastrophe." *transversal*, March. www.eipcp.net/transversal/0608/solomon/en. Originally published in *Borders, Nations, Translations: Übersetzung in einer globalisierten Welt*, edited by Boris Buden, 165–78. Vienna: Turia and Kant, 2008.

Solomon, Jon. 2010. "The Experience of Culture: Limits and Openings of Foucault's Eurocentrism." *Transeuropeénes: Revue internationale de pensée critique*, no. 1. http://www.transeuropeennes.eu/en/articles/108/The_Experience_of_Culture_Eurocentric_Limits_and_Openings_in_Foucault

Solomon, Jon. 2014. "The Postimperial Etiquette and the Affective Structure of Area." *Translation* 4: 171–201.

Tanaka, Stefan. 1995. *Japan's Orient: Turning Pasts into History*. Berkeley: University of California Press.

Terranova, Tiziana. 2004. *Network Culture: Politics for the Information Age*. Ann Arbor, MI: Pluto Press.

Venn, Couze. 2009. "Neoliberal Political Economy, Biopolitics, and Colonialism: A Transcolonial Genealogy of Inequality." *Theory, Culture, and Society* 26, no. 6: 206–33.

Virno, Paolo. 2004. *A Grammar of the Multitude: For an Analysis of Contemporary Forms of Life*, translated by Isabelle Bertoletti, James Cascaito, and Andrea Casson. New York: Semiotext(e).

Walker, Gavin. 2011. "Primitive Accumulation and the Formation of Difference: On Marx and Schmitt." *Rethinking Marxism* 23, no. 3: 383–404.

Wolfe, Cary. 2010. *What Is Posthumanism?* Minneapolis: University of Minnesota Press.

Liquid Area Studies:

Northeast Asia in Motion as Viewed from Mount Geumgang

Tessa Morris-Suzuki

Prelude: Mt. Geumgang

In July 2008 a South Korean woman named Park Wang-Ja was shot dead by North Korean security forces as she walked on a beach at the Mt. Geumgang tourism complex, which lies just on the north side of the line dividing the Korean Peninsula. Her killing sparked a crisis in North-South Korean relations. Since 2000, Mt. Geumgang (also transliterated as Mt. Kumgang, known as Mt. Kongō in Japanese, and in fact constituting a small mountain range known as the Diamond Mountains in English) had been the site of a major joint tourism development run by both the North Korean authorities and the South Korean Hyundai conglomerate, but the death of Park Wang-Ja led to the closure of the tourist resort, a significant step in a rolling back of the joint North-South projects that had emerged from the Sunshine

positions 27:1 DOI 10.1215/10679847-7251897
Copyright 2019 by Duke University Press

Policy of the Kim Dae-Jung and Roh Moo-Hyun eras. Despite negotiations on reopening, the project remained in limbo from the middle of 2009 to the start of 2018, when renewed dialogues between North and South Korea raised hopes that it might be revived. The incident at Mt. Geumgang was extensively reported in the international media. Few of the reports, however, mentioned anything about Mt. Geumgang's long and fascinating past, and I could not help wondering how many viewers of news items on the shooting were aware just how deeply Mt. Geumgang is embedded in regional history.

In this essay, I use the example of the history of Mt. Geumgang as a starting point for exploring a regional perspective on Northeast Asia. I might have chosen any one of a number of similar sites as a starting point, for my underlying point is simply this: by selecting a particular place, and by exploring the changing network of paths through which it has been linked to the surrounding world, we can learn interesting things about whole regions, about the way in which those regions themselves have changed over time, and about the very concept of "region" or "area" itself. The point on the map from which we choose to view a region may be a capital city or a major trading port. But places of religious, artistic, and touristic pilgrimage like Mt. Geumgang also provide valuable vantage points for observing the changing flows of people and ideas that constitute the region. This essay, then, is an attempt to say something about the nature of Northeast Asia as viewed from Mt. Geumgang.

Northeast Asia: The Incomplete Region

In 2003, US social scientist Gilbert Rozman published a book entitled *Northeast Asia's Stunted Regionalism* (2003). The book offers a careful analysis of factors hindering the process of Northeast Asian political and economic integration since the 1990s. Its title, however, expresses a perception of Northeast Asia which is widely repeated in writings both within and outside the region. By contrast with Western Europe—widely seen as providing a model of integration based on shared history and values—Northeast Asia is frequently depicted as being an incomplete, unsatisfactory, somehow disappointing region.

In Western Europe since the end of World War II "the growth of regional identity has been associated with institution building and confidence building" (Wright 2006: 172). But more than seventy years after the end of the Asia-Pacific War, Northeast Asia (as we are often reminded) seems almost as deeply divided as ever (Qi 2005: 1–49). The comparison with Southeast Asia, too, is often an unfavorable one. While Southeast Asian nations have found common cause in the Association of Southeast Asian Nations, Northeast Asia remains fragmented particularly by the forces of nationalism.[1] The persistence of nationalism in Northeast Asia is indeed one of the reasons why, as Kent Calder (2001: 106–21) has written, the region is widely seen as being "among the most dangerous places on earth." From this perspective, Mt. Geumgang would surely be seen as symbolic of everything that is dangerous about Northeast Asia: located next to the world's only remaining Cold War border, it is a flash point in the still-unresolved relationship between the two halves of the divided Korea—and this volatile relationship has the power to shape the destiny of the entire region.

Against this background, much of the recent debate on Northeast Asian regionalism has focused on seeking explanations for the failure, or at least the weakness, of regional integration. While some scholars (such as Rozman) seek these explanations in recent political and economic trends, others see them as lying more deeply embedded in history and culture. For example, Kim Soung Chul (2008: 226) of South Korea's Sejong Research Center argues that "the unique historical experiences of Northeast Asia have resulted in a curious mix of certain nineteenth century pre-modern national aspects, twentieth century balance of power conflicts, and twenty-first century post-modern, multilateral order. This peculiarity will make it more difficult for Northeast Asia to form a regional community like Europe."

Rather than entering into the debate about the sources of Northeast Asia's incomplete regionalism, I will reverse its logic. In other words, rather than asking why Northeast Asia has remained a "stunted region," I will use an example from the history of Northeast Asia for questioning the way in which area studies conceptualizes the very idea of a region or area. Much has been written over the past two decades about area studies as an approach to social knowledge. Debates about the value and limitations of the area studies approach have often drawn empirical examples from Southeast Asia. A fresh look at Northeast Asia as area can, I believe, provide some useful insights into the possibilities and limitations of area studies in the twenty-first century.

Area Studies and Beyond

Area studies flourished in the universities of the United States and its allies during the Cold War decades from the 1950s to the 1980s. Since the 1980s, however, the area studies approach has faced serious challenges from various directions. According to some critics, the creation of interdisciplinary academic units focusing on particular regions of the world (such as Southeast Asia, Africa, or Latin America) has led to an erosion of the disciplinary basis of academic research, and so to a loss of scholarly rigor (see Anderson 1978).

A second source of criticism has been the fact that, particularly in the United States, area studies became closely connected to Cold War strategy, with substantial amounts of funding coming from government agencies involved in national strategic planning. As a result, the agenda of area studies (it is argued) is always in danger of being captured by the state (see, for example, Wallerstein 1998). The urge to harness area studies to US strategic objectives did not, of course, disappear with the end of the Cold War. Twenty-first century publications by the US Army War College, for example, vividly illustrate how this tradition survives. One paper on Northeast Asia produced by the college in 2004 emphasizes the need to understand the cultures of Northeast Asian countries so that the United States (and particularly the US military) may "continue to maneuver successfully to maintain and sustain our interests" in the region (Rogers 2004: v).

To assist in this effort, the author goes on to explain that unlike the people of the West, for whom "the scientific method has always been our means of discovery and influenced our behavior at the most basic levels," the people of Northeast Asia have a culture that was "not influenced by the scientific method until . . . recently" (4). While Westerners are individualistic (we are told), Northeast Asians are group oriented, thus, "approach a NEA [Northeast Asian] leader and you approach an entire culture" (7). In dealing with a country like the People's Republic of China, the author therefore stresses that "we cannot use logic or causality statements to predict or influence their actions. They are experientially based, and action to consequence logic doesn't carry weight. Approaching China experientially and leveraging their view that nature will produce the right result over time will improve the chances of our success" (11–12). Studies like this illustrate not only the ongoing connection between cultural research and military strategy but also the remarkable durability of 1950s-style, essentializing, homogenized views of culture that gave area studies such a bad name.

In most parts of academia, though, area studies has moved on well beyond such sweeping generalizations about Northeast Asians' lack of individualism or scientific rationality. Influenced by postcolonial and cultural studies, area research has become more self-reflexive and questioning about its own methods (see, for example, Jackson 2003). Rather than being a field of scholarship through which the hegemonic West observes and analyzes the non-Western "Other," it has increasingly become a transnational endeavor, in which multiple voices and perspectives have a place.

Yet profound conceptual challenges remain. The most profound concerns the nature of the area itself—in other words, the connection between human society and geographical space. Conventionally, area studies has been based on a rather static perception of areas as spaces contained by physical geography and environment, and defined by shared culture. In other words, geographical conditions—plains, mountain ranges, rivers, seas—were assumed to bring certain groups of people into close contact with one another, while separating them from other groups. Within these shared landscapes, over millennia, people developed cultural patterns that were adapted to their natural environment. As the French historian Fernand Braudel (1995: 10) put it, "to discuss civilization is to discuss space, land and its contours, climate, vegetations, animal species and natural or other advantages. It is also to discuss what humanity has made of these basic conditions." Area studies focused on these cultural or civilization areas, bringing together insights from the disciplines of history, geography, anthropology, linguistics, sociology, and so on, in an effort to identify the fundamental characteristics of the shared regional culture, and its impact on the social, economic, and political destiny of the people of the region. In the immediate postwar decades, large amounts of effort went into the task of trying to identify the underlying patterns of value, belief and thought that characterized each cultural area, for it was widely accepted that "a society's reactions to the events of the day . . . are less a matter of logic or even self-interest than the response to an unexpressed and often inexpressible compulsion arising from the collective unconscious" (22).

More recent research, however, has raised fundamental questions about this view of space, society, and culture. The areas of area studies, for example, are normally continents, subcontinents, or other contiguous blocks of land, such as Europe, South Asia, Africa, and so forth. Arjun Appadurai and others, on the contrary, have sought to develop a form of area studies based not on visions of great blocks of shared territory or "civilizations" but on the social and cultural "precipitates of various kinds of action, interaction and motion—trade, travel, pilgrimage, warfare, proselytization, colonization, exile, and the like" (Appadurai 2001: 7-8). The outcome of this shift, as Willem van Schendel has argued, is a vision of region as shifting and polymorphous, capable of assuming "unfamiliar spatial forms—lattices, archipelagoes, hollow rings, lattices" (Schendel 2002: 664). One outstanding example of such unfamiliar forms, of course, is Schendel's notion of "Zomia," a region that cuts across national boundaries and traditionally-defined regional boundaries and links the upland areas of inland Southeast Asia and southern China: a notion that has in turn been further developed by James Scott in his study *The Art of not Being Governed* (Scott 2009).

Meanwhile, as the work of scholars such as Hamashita Takeshi (1999) and Barbara Watson Andaya (2007) has shown, the conventional focus on areas as blocks of territory neglects the enormously important human connections created by seas and oceans. Historically, travel in many parts of the world has been easier by sea than by land. The result has been the creation of long trading routes, linking people in fishing villages or port towns that may be separated from one another by great distances. When we look again at these neglected maritime connections, we are forced to rethink static and essentialized views of regional culture. For the people of trading ports, while sharing some cultural commonalities with the people of the hinterland, may also share other cultural features with trading partners situated far away on the other side of the ocean.

Though the maritime world is particularly rich in such trading routes, other pathways of travel also cross continents and deserts, creating both economic links and cultural commonalities between people from widely dispersed communities. The various Central Asian silk routes that appeared and disappeared between ancient times and the sixteenth century are examples of such inland pathways, as are the routes of religious pilgrimage that traverse regions and continents.

Mt. Geumgang, which is a relatively inaccessible rugged area on the east coast of the Korean Peninsula, is an example of influence of such pilgrimage routes. Its soaring mountains and spectacular landscape made it a favored site for the building of Buddhist temples, and for retreat from the world for the purpose of meditation or artistic creation. It is not entirely clear when the first Buddhist monks came to the mountains, but the earliest temples and hermitages to be constructed there date from the Silla period (57 BC to 935 AD) and include Singye-Sa (CE 519) and Jangan-Sa (CE 551) (Jeong 1992: 256). Relatively soon after the introduction of Buddhism to Korea, the mountains had become a center for pilgrimage, renowned not only throughout the Korean Peninsula but also in China (see Yu 1998: 11–38, esp. 29–30). The Yuan Emperor was said to have contributed to the rebuilding of Jangan-Sa, the greatest of the mountains' temples, in 1348, and in the middle of the twentieth century the nearby temple of Yujeom-Sa contained fifty statues of the Buddha, some of which statuary had reached the temple ten centuries earlier from Tokharistan (a central Asian kingdom on the borders of modern day Uzbekistan, Tajikistan and Afghanistan) (Tokuda 1918).

Closer attention to such trade and travel routes creates a more dynamic and fluid view of the relationship between society, culture, and space. Rather than being seen as embedded in geography, and thus constant over millennia, cultural areas come to be seen as dynamic and overlapping, constantly created and re-created by human movement and interaction. The result is a view of space influenced less by the worldview of Fernand Braudel than by that of his contemporary Henri Lefebvre. Lefebvre, of course, is remembered for his profound reflections on "the production of space," in which he emphasized the fact that social space is something that does not exist of itself but is always *created* by social practice. In the modern world, however, the practices of capitalism, while creating particular forms of space, also work to persuade us that these spaces are natural—to conceal the fact that space is the product of human action. Space, in other words, is both constructed and profoundly ideological (Lefebvre 1991).

This deepening sense of the construction of space is also linked to a growing scholarly interest in human mobility. While studies of society have traditionally tended to see stasis as the normal state of human beings, there is now a growing wealth of research that emphasizes that human societies are themselves constituted by motion (both within and across national borders). In his work on "migration as method" for example, Iyotani Toshio has questioned the view of migration as an exception to the norm of settled existence, and thus above all as a focus for control by state authorities (Iyotani 2005). We might, he suggests, equally well reverse the question commonly posed by migration researchers—what is it that impels people to migrate?—and instead ask, what are the special conditions that impel some groups of people to stay still?

In an earlier essay (see Morris-Suzuki 2000), I discussed the limitations imposed by the conventional geographical image of cultural or civilization area and argued the need for an "anti–area studies" that would take widely dispersed and differing points on the face of the globe (points that may be cities, minority communities, frontier zones, etc.) as a basis for exploring shared historical and social issues. In this way, I suggested, it might become possible to develop new perspectives on the multiple and overlapping social spaces that humans inhabit, and to overcome the dichotomized sense of space that locates the area studies specialist as exterior to the area that he or she studies.

In this essay, I want to extend these arguments by showing, through a simple practical example, how a point on the face of the globe becomes, over the course of history, woven into multilayered and ever-changing networks of connection—multiform and protean areas whose shifting dimensions can give us new insights into the connections between geographical space and social and cultural history.

Regions, Flows, and Vortices

To describe this venture, we might use the term *liquid area studies*. The starting point of liquid area studies is the notion that an area is *brought into being only by human activity*—travel, trade, and communication. Through interacting with one another over relatively large distances, people discover or create commonalities of lifestyle or understanding. Some of these commonalities are newly created, but some emerge from the rediscovery or reinvention of connections that existed in the past.

In other words, rather than being a solid thing, embedded in the bedrock of geography, an area is rather like a fountain, which is given shape only by constant activity and movement. Like a fountain too, it may radically change shape, or disappear altogether, if movement changes direction or ceases. Several consequences follow from this view of areas as constituted by human action and movement. The first is that an area may take a wide variety of shapes. It may link together people living in a large, contiguous block of land, but it may also be made up of a series of linked dots far removed from one another in space. A second consequence is that areas may overlap. That is to say, because people live close to one another does not necessarily mean that they trade, communicate, and share ideas with one another. Within the same town or rural area it is possible to have different groups of people who live in different social spaces and thus participate in different areas. Third, liquid area studies offers no grounds for assuming that the cultural cohesion and integration of any area will survive unchanged over long periods of time. Although in some places stable patterns of interaction may result in the creation of a rather long-lasting area, elsewhere, the contours and nature of the area may be very fluid, undergoing repeated and dramatic metamorphoses over time.

A liquid area studies approach implies that, rather than beginning with a search for the geographical boundaries of the region and for the environmental conditions that shape its culture, it may be more useful to start by turning our attention to two dimensions of human interaction: flows and vortices. Flows, obviously enough, are movements of people, goods, and ideas that link social groups together. They may be migratory movements, trade routes, paths of pilgrimage, or trajectories traced and retraced by advancing and retreating armies. A social and cultural area is constituted by a particularly rich intersection of such flows.

A vortex, on the other hand, is a place where multiple flows meet, creating a whirlpool of swirling social and cultural interaction. This may be a city or an island at the intersection of multiple trade routes, a strategically significant point where contending political forces come into repeated contact and conflict, or a place of pilgrimage that brings together the faithful of many backgrounds. Like the points where sea currents collide, such vortices are often particularly rich in the nutrients that sustain life or (in this case) sustain the seeds of social and cultural change. However, just as flows shift over the course of history, so do vortices. Places that have been vital meeting points or trading posts wither into insignificance as paths of migration and commerce shift, leaving them high and dry, and new flows and vortices appear elsewhere.

In using the term *liquid area studies*, however, I seek to suggest not simply the need to conceive of the geographical space we study as "liquid" but also to "liquefy" the process of area studies itself. The traditional model of area studies posited the existence of a stable geographical area, which was the object of scrutiny by an academic expert whose place and identity were also firmly defined and were normally located outside the field of study. Liquid area studies, on the contrary, implies not only that the shape of the area is protean but also that the locus of the researcher is shifting and unstable. The observer may at times exist outside and at times be within the field that he or she studies.

The Changing Shape of Northeast Asia—The Sino-Centric World

The very characteristics that, from the perspective of many area studies scholars, make Northeast Asia a "stunted" or "incomplete" region may, from a liquid area studies perspective, provide a valuable starting point for rethinking the protean nature of social and historical space. Over the course of centuries, the routes and connections that constitute the space we call *Northeast Asia* have passed through phases of integration and disintegration, each time emerging with transformed contours and dynamics. Repeated phases of regional conflict and fragmentation have severed the flows that bind the region together. Renewed phases of integration have required reconnecting old linkages and building new ones. The region is thus forever being reconstructed and reinvented. One way of tracing these waves of reinvention is to consider how the flows that converged at a particular vortex have changed over time.

The Sino-centric Northeast Asia of the Ming (1368–1644) and early Qing (mid-seventeenth to eighteenth centuries) Dynasties was held together by a network of trade, tribute, and communication routes, not only linking outlying parts of the region to the Chinese imperial capital but also connecting cities such as Edo and Hanyang/Seoul, and trading posts like Qiqihar in Manchuria to their hinterlands. The structure of these flows also extended to include areas such as Annam, the northern part of what is now Vietnam, and its capital Dōng Kinh (today's Hanoi), which, for the purposes of present-day area studies, are more commonly seen as part of the Southeast Asian region.

At the peak of the region's integration, this network of travel routes was dense and complex. Tribute missions from Korea traveled up the Korean Peninsula and through Manchuria to the Ming capital of Beijing with increasing frequency—a sign, not of growing Korean dependence on China, but rather of the fact that tribute missions were profitable ventures for those who took part in them. China maintained complex and multiethnic intelligence networks extending to the Ryūkyū Kingdom, Annam, and beyond (Tsai 1995). Many of the empire's connections to the world of maritime Asia, meanwhile, flowed through eastern ports like Ningbo, which was closely linked to port cities both of the southern part of Korea and of southwestern Japan (Hamashita 2003: 17–50, esp. 19; Levenson 1991: 386).

Routes of trade and diplomacy often converged or intersected with routes of religious pilgrimage. The spread of Buddhism throughout Northeast Asia, superimposed on earlier Taoist and shamanistic beliefs, created paths of pilgrimage that, for example, linked the four major Buddhist sacred mountains of China westward to sites like the Mahabodhi Temple in Northwestern India and eastward to such places as Mount Geumgang on the Korean Peninsula. As early as the eighth century, the Korean monk Hye-Cho had traveled from the Kingdom of Silla to India and Central Asia, leaving a written record that includes descriptions of Tokharistan, Bamiyan (in today's

Afghanistan), the "Land of the Arabs," and parts of the Byzantine Empire (Hye Ch'o 1985). Travel along these routes was, of course, multidirectional: the Indian monk Dhyanabhadra (a key figure in the fourteenth-century development of Chan/Seon Buddhism) was among the many pilgrims who paid visits to Mt. Geumgang (Grayson 2002: 98–99).

Increased interaction between Japan and the Asian continent during the Ming Dynasty promoted religious, intellectual, and artistic interaction, making it possible, for example, for a Japanese monk like the renowned brush painter Shūbun (ca.1414–63) to travel to Korea in search of Buddhist sutras, and for his Korean associate and namesake Yi Jumun (Ri Shūbun) to travel to Japan. It is not clear whether the Japanese Shūbun visited Mt. Geumgang, but he surely knew of its existence, and its imagined if not real presence influenced the visions of landscape that both he and his Korean counterpart brought to Japan. During his stay in Korean temples, Shūbun would have encountered religious and artistic ideas from China, brought back by Korean monks and emissaries who traveled northward to Beijing or southward on trading vessels to Ningbo (Levenson 1991: 386). A good example of this interaction is the itinerary traversed by Shūbun's most famous disciple Sesshū (1420–1506), who in the 1460s took advantage of close trade connections between Yamaguchi and Ningbo to travel via Ningbo to Beijing, from where he visited Tianjin, Nanjing, Souzhou, and other parts of China before returning to Japan more than two years later with a wealth of ideas that were to exert a profound influence on the development of Japanese landscape painting.[2]

Cities such as Beijing and ports like Ningbo were thus vortices, where flows from many directions came into contact. In Beijing, for example, a traveling monk and artist like Sesshū would have met not only Chinese but also Korean envoys and artists, as well as (perhaps) travelers from Annam or Central Asia. Mt. Geumgang was also, in its small way, a vortex in these regional flows. Its monks traveled, both on pilgrimages to other famous sites in Korea and China, and also on journeys to seek alms for the upkeep of their monasteries.[3] As a place of pilgrimage, meanwhile, its temples attracted not only devout Buddhists but also prominent Korean officials like Yi Guk (1298–1351) and Gweon Geun (1352–1409).[4] As well as traveling to Beijing on diplomatic missions, Gweon also spent time at Mt. Geumgang, composing literary works including a poem written there in 1396, just after the establishment of the Joseon Dynasty, in which he describes himself amid the "protruding and precipitous mountain peaks," looking down on "the chaotic state of heaven and earth" (Gweon 1936).[5]

Though the Joseon Dynasty is generally known as a period of state hostility toward Buddhism, these religious and intellectual journeys continued, and by the middle of the dynasty they were increasingly taking on some features of modern tourism, with the fame of a place like Mt. Geumgang being broadcast by the works of Korean literati who wrote travel accounts of their visits to the mountains, and by the many visual images of the Geumgang mountain range created by famous artists like Jeong Seon (1676–1759, also known as Gyeomjae) and Kim Hong-Do (1745–1806, also known as Danwon) (Kim Dong-Ju 1999).[6] While monks and other "professional" pilgrims traveled great distances, the popular pilgrimages of the seventeenth, eighteenth, and nineteenth centuries drew visitors from nearby towns and villages into smaller circuits of movement that intersected with the great circuits of the traveling monks. One account from seventeenth-century China by the scholar and official Chang Tai describes the popular pilgrimage to the Buddhist complex of Mount Tai as follows:

> Each day there are several thousand visitors, who will occupy hundreds of rooms and consume hundreds of vegetarian and ordinary banquets; they are entertained by hundreds of actors, singers and musicians, and there are hundreds of attendants at their beck and call. The guides are from about a dozen families. On an average day eight thousand to nine thousand visitors come, while the number can reach twenty thousand on the first day of spring. (Pei-Yi Wu 1992: 74)

The British imperial statesman Lord Curzon, visiting Mt. Geumgang toward the end of the nineteenth century, described the large number of local Korean visitors to the mountains as "equivalent to the English Bank Holiday young man on a bicycle—a character very common among the Koreans, who cultivate a keen eye for scenery, and who love nothing better than a *kukyeng* [*gugyeong*], or pleasure trip in the country" (Curzon 1894: 105).

Between the time of Chang Tai and the time of Lord Curzon, however, the disturbing effects of the growing intrusion of Westerners into Asia had produced reactions that in some cases restricted the flows of movement that had crossed the region, and from the late sixteenth century on other developments also disrupted existing networks. The mountain ranges around Mount Geumgang had constituted a physical watershed in a historical event that helped to determine the shift toward a more fragmented region, for it was here in 1593 that the march of Japanese ruler Toyotomi Hideyoshi's forces through the Joseon Kingdom toward the Chinese empire encountered one of their most severe setbacks, at the hands of a force that included warrior monks recruited by two of the central figures in the development of Seon Buddhism in Mt. Geumgang—Seosan and (1520–1604) and Samyeong (1544–1601).[7]

The great social disruption caused by the Japanese incursion, followed by the Manchu invasion of the Korean Peninsula and the establishment in 1644 of the Manchu Qing Dynasty in China, severed many of the routes that bound the Japanese Archipelago to the Korean Peninsula, and the Korean Peninsula to China. Although tributary relations between Korea and the Chinese Emperor were restored, and trade between Japan and Korea via the island of Tsushima was revived, mobility within the Northeast Asian region was more tightly circumscribed from the seventeenth to the mid-nineteenth centuries. Religious pilgrimages continued to flourish but were increasingly constrained within narrower geographical spheres. With the declining power of the Qing Empire, the network of flows that had sustained Sino-centric Northeast Asia gradually atrophied.

From the perspective of Mount Geumgang, then, the region was not a stable geographical block made up of the countries that we now know as China, Korea, and Japan. Rather, it was a set of intersecting circuits and pathways. For many of the fishermen and farmers of the coastal strip at the foot of the mountains, the sphere of movement and contact with the outside world was very narrow—confined perhaps to a handful of villages. The more prosperous, however, would have joined visitors from further afield—from the local administrative center of Tongcheon and even from as far afield as Wonsan and Seoul—in visits to the temple complex. And the proximity of the complex would from time to time also have given them at least a glimpse of people and objects that reached their world along an enormously extended web of links, connecting this Mount Geumgang to the Chinese Ming Dynasty and Korean Goryo and Joseon Dynasty capitals, to the four Buddhist holy mountains of China (Mount Wutai in the North, Mount Omei in the West, Mount Jiuhua in the South and Mount Puto, on an island off the east coast), to port cities like Ningbo and to the major Buddhist monasteries of India. Such networks, however, waxed and waned in response to political and social change, creating a fluid and ever-changing lived experience of region.

A focus on spatial flows tends to make us think in horizontal terms of movement over a flat surface, but it is important to emphasize that historically, as in today's globalized world, flows through space were inextricably imbued with vertical inequalities of power. Zygmunt Bauman's (1998: 85) aphorism "divided we move" was at least as applicable to fourteenth century Mt. Geumgang as to the world of the twenty-first century. Not only were the more powerful more likely to be mobile, in a very real sense they traveled on the backs of the less mobile. One of the earliest Western visitors to the mountains noted the Joseon Dynasty practice of requiring local communities to shoulder the cost of travel by members of the official class (*yangban*): each village was expected to provide "food, lodging, money, bearers, and beasts of burden on the spur of the moment whenever it is so unlucky as to have an official visitor" (Campbell 1891: 5).

Imperialism and Reintegration from the 1880s to the Mid-Twentieth Century

After the narrowing of the arteries of contact during the preceding centuries, from the late nineteenth century onward a radically transformed set of flows emerged: flows profoundly shaped by the Western presence, by the rise of nationalist movements in the region and above all by the force of Japanese imperial expansionism. The region was reintegrated under Japanese hegemony: the new, Japan-centered Northeast Asia of the early twentieth century is indeed a reminder of the fact that integration, far from being an inherently benign process, may involve both violence and hierarchies of power.

Japan's dominance in Northeast Asia was confirmed by its victory in the two wars that are commonly known as Sino-Japanese War (1894–95) and the Russo-Japanese War (1904–05), but which (I have argued elsewhere) might best be seen as parts of a single ongoing conflict that could be called "the First Korean War" (since competition for dominance over the Korean Peninsula lay at the root of the conflict, and much of the fighting also took place in Korea). The new Northeast Asian networks of mobility that emerged from Japan's victory in this First Korean War were very different from those of the earlier, Sino-centric order. Their center of gravity had moved east, while southern areas such as Annam were now incorporated into the colonial world of French-controlled Indochina. Key cities like Beijing had declined in influence while others—including Osaka, Shanghai, and Dalian emerged as the new vortices of the reshaped region.

The flows of people, goods, and ideas that formed this modern Northeast Asia moved along a new network of infrastructure, centered on the railways, which were among the major prizes over which the First Korean War had been fought: the Chōsen Railway (started by Korean engineers but completed by Japanese colonizers) and the South Manchurian Railway, with its various branches. The Chōsen Railway, running northward through Keijo (Seoul), Pyongyang, and Sinuiju, roughly followed the path taken in earlier times by tribute missions to the Chinese empire, but was linked into a new set of connections at either end, and embodied a wholly new set of power relationships. By the time of World War I Japanese transport planners already dreamed of an even greater network of routes that, via tunnels on both sides of the world, would one day connect Tokyo to London directly by rail. World War I and the Russian Revolution helped to ensure that the dream was never realized, but the Chōsen and South Manchurian Railways became arteries conveying nutrients through a new Northeast Asian geobody.

The railway was, of course, not merely metal and engineering, stations and timetables, but also political power. As one employee of the South Manchurian Railway Company put it, paraphrasing the company's first director Gotō Shinpei, "Japanese imperialism in its advance into Manchuria . . . chose to assume the form of a railroad company" (Itô 1988: 5). The railway carried Japanese businessmen, soldiers, and colonial settlers throughout the continent, as well as Chinese and Korean migrants into Manchuria. It carried Japanese artists and literati who traveled through Korea and China, and Korean and Chinese scholars who traveled to Japan for study or work. It was, then, an instrument both of violence and of peaceable exchange of ideas. Its network carried Japanese police engaged in the suppression of Korean nationalism, but also Korean nationalists who used the trains to move back and forth (sometimes in disguise) between Korea and Manchuria. For, as Park Hyun-Ok (2005) has vividly shown in her book *Two Dreams in One Bed: Empire, Social Life, and the Origins of the North Korean Revolution in Manchuria*, the new network of flows created by imperialism was inventively used and expanded by a wide range of groups, both rulers and ruled.

The microcosm of Mt. Geumgang illustrates some features of these new regional dynamics. After a period of relative quiescence in the nineteenth century, during the first half of the twentieth century, its temples became the focus of a revived flow of travelers—this time, of secular pilgrims. The origins of these visitors, and the paths by which they reached the mountains, had changed since the heyday of the fourteenth and fifteenth century pilgrimages. After Japan's formal annexation of Korea in 1910, a new ferry route was established, linking Mt. Geumgang (Kongō-san) to the port city of Wonsan, and in 1931 the completion of the Mt. Kongō Electric Railway[8] linked Mt. Geumgang into the Japanese-controlled Korean and South Manchurian rail networks. The development of these transport links was part of a colonial policy that singled out Mt. Geumgang as a site on the tourist trail through Korea and Manchuria, energetically advertised to Japanese and foreign travelers alike. Between 1925 and 1938, the annual number of visitors to Mt. Geumgang rose from less than 200 to over 24,000 (Kongōsan 1939: 70).

Most of these visitors came either from within Korea or from Japan, but some also came from the United States by ship or from Europe via the Trans-Siberian Railway. Quick to grasp the propaganda value of tourism, the Japanese authorities created tours carefully designed to introduce foreign visitors to the triumphs of the Japanese colonial venture: "Korea of yesterday, Chosen of today invites you," proclaims one Chōsen Government Railways English-language advertisement, offering "Kongō-san, Chosen's Diamond

Mountains—the future vacation land of the Far East" as the highlight of a tour through "a land in the making—an ancient land that year by year is losing its garments of old and donning the garb of youth."[9]

For Japan's artists and literati who encountered the mountains in the first half of the twentieth century, Mt. Geumgang acquired a place not unlike the position occupied by the Swiss Alps in the nineteenth-century European romantic imagination: a point reinforced by the Swiss-style chalets constructed as hotels by Japanese entrepreneurs in the mountains. The pioneering newspaper editor and social commentator Tokutomi Sohō (1863–1957) composed poems extolling Geumgang's peerless beauty (Mantetsu 1924: 2), while Ishii Hakutei (1882–1958), Maruyama Banka (1867–1942), and a host of other early twentieth-century landscape artists traveled from Japan to the mountain range to paint its astonishing landscape. By the 1930s plans were underway in Japan to have Mt. Geumgang declared a national park, alongside sites such as the Seto Inland Sea, Mt. Unzen, and the Fuji Hakone district, all of which had been designated under the new National Parks Law of 1931; but the war intervened, and the plans came to nothing (165–66; see also Brovko and Fomina 2008: 211–25; Ministry of the Environment 2017).

The region into which Mt. Geumgang was now reincorporated was thus constituted of flows very different from those of earlier eras. Physically, they brought visitors across the East Sea from Japan, across the Pacific from America and across Siberia from Europe, while the objects, ideas, and imaginings that visitors brought with them came no longer from the Himalayas and the sacred mountains of China but from the famous mountains of Japan and the Western world. This reconnection of Mt. Geumgang to the new flows of Japan-centered Northeast Asia involved both a rediscovery and a forgetting of the past. A major attraction of the region for tourists was its wealth of ancient Buddhist temples, hermitages, statues, and rock carvings, created by earlier generations of pilgrims. But these physical objects were often presented with little historical content—as works of art created in the mists of history and then allowed to decay until touched by the restoring hands of colonizers. Although a few Japanese visitors to the mountains evoked the works of Joseon Dynasty poets and artists (Tokuda 1918), most Japanese tourist literature was silent about this heritage: one publication even (remarkably) described the British traveler Isabella Bird, who visited

Mt. Geumgang in the 1890s, as "the first explorer of Kongō-san, and in a sense its discoverer" (Department of Railways 1920)—as though all the myriad of earlier depictions of Mt. Geumgang by Korean travelers, poets, and artists had never existed.

The erasure of memory went hand in hand with material exploitation, particularly of the large numbers of Korean laborers who were employed to build the railway, or to work as "coolies" carrying the baggage of tourists, and sometimes actually carrying the tourists themselves, up the precipitous slopes of the mountains. On the other hand, as Park Hyon-Ok shows in her discussion of the interaction between Chinese, Korean, and Japanese communities in Manchuria, systems created by the colonizing power could not be wholly controlled by the colonizers. Korean tourists, including growing numbers of school groups, also took advantage of the rail links and hotels created by Japanese colonialism to rediscover their own history and culture. The excitement of this rediscovery is vividly illustrated, for example, by an essay written for a student magazine in the 1938 by Kim Ok-Seon, a young woman student from the town of Hamheung, who traveled on the Mt. Kongō Electrical Railway to visit the Buddhist sites of Mt. Geumgang with her teacher and classmates (Kim Ok-Seon [1938] 2007: 29–38). Another role of the tourist destination as multinational meeting point is illustrated by the memoirs of the US journalist Helen Foster Snow (who also wrote under the pseudonym Nym Wales). Snow (Snow 1984: 186) recalls sitting on the veranda of the Japanese-run Uchi-Kongō Hotel (at that time Mount Geumgang's best hotel) in 1936 with "close-mouthed missionaries who hated Japan and felt a special responsibility for Korea, the only Protestant community of importance in Asia" (Snow 1984: 186), an encounter that led to her later contacts with Korean independence fighters, about whom she wrote a sympathetic account in her book *The Song of Ariran*.

These varied travel experiences illustrate an important feature of the imperial multiculturalism of Northeast Asia from the 1910s to the early 1940s. The new flows of the modern region created ethnically diverse cities like Shanghai, Mukden (today's Shenyang), and Andong (Dandong), with their communities of Chinese, Japanese, Koreans, Russians, and Western Europeans. But, like oil and water, the various groups flowed side by side while only rarely intermingling. So Japanese travelers like Ishii Hakutei or poet

Ōmachi Keigetsu could move through Manchuria and Korea visiting Japanese notables, staying in Japanese hotels, and existing in a world which, but for the surrounding scenery, was almost entirely "Japan."

The Cold War Decades, 1945–1980s

The collapse of the Japanese empire in 1945 brought with it an abrupt disintegration of the links that had connected Northeast Asia throughout the first decades of the twentieth century. The forces of division that tore the region apart were, indeed, far more complex and profound than those that divided postwar Europe. Europe was divided by a single "Iron Curtain" separating West from East, but on the western side of the Curtain cross-border movement was relatively easy and cultural communication flourished. Though bifurcated by this Cold War divide, Europe did not experience the "hot war" of armed conflict between the new superpowers, as Northeast Asia did during the Second Korean War (1950–53).

In Northeast Asia, the 38th parallel became just one of many Cold War dividing lines, although it has proved the most enduring. The Sino-Soviet split of 1960 created a divide right across the region, and even in the noncommunist parts of Northeast Asia, the Cold War order created borders that isolated national societies and divided families. In China, North Korea, and the eastern parts of the Soviet Union, communist governments imposed strict controls on human movement both within and across their borders. Under the US military umbrella, meanwhile, efforts to contain any threats of "communist subversion" included tight restrictions on the movement of people between South Korea, Taiwan, and Japan, and even between Okinawa and Japan. But, while mobility within the region was curtailed, dense flows of goods, ideas, and people between the United States and each of the US allies in the region flourished. Indeed, the network of US bases in South Korea, Japan, and Okinawa became an archipelago through which American military forces flowed in a constant stream unchecked by national border controls. The vision of Northeast Asia as a region irrevocably divided on national lines need to be understood against the background of this Cold War system.

The drastic sundering of the human and material flows that constituted the region is vividly evident from the Cold War destiny of Mt. Geumgang. Links not only to Europe and North America but also to Japan and to the southern half of the divided Korean Peninsula were suddenly severed. Mt. Geumgang experienced some of the most intense fighting of the Korean War. The temple of Jangan-Sa was converted into a North Korean prisoner-of-war camp, and was later totally destroyed, as was nearby Yujeom-Sa. The mountains were suddenly isolated, not just from Tokyo and Seoul, but even from villages like Hwajipo, barely 15 kilometers away to the south. The landscape and artistic treasures of the mountains (at least, those which had survived the war) remained a place of secular pilgrimage, but now the vast majority of the visitors came from within the borders of the Democratic People's Republic of Korea. Ideology reshaped space.

During the 1960s workers' rest centers were developed in the mountains, and ordinary people as well as party cadres visited the Mt. Geumgang to revive their spirits in its landscape and fresh air. One refugee from North Korea to whom I have spoken still remembers with pleasure the good food and breathtaking scenery that he experienced on his holiday at Mt. Geumgang in 1966. He traveled there from his factory in the northern border city of Sinuiju, and was afforded a fleeting glimpse of that easier life that Kim Il-Sung's regime promised the North Korean people but ultimately failed to deliver. (The rest center at Mt. Geumgang apparently ceased to be accessible to ordinary workers after about the end of the 1960s) (pers. comm., Seoul, June 20, 2005).

The rugged mountain landscape now became a symbol of Korea's unyielding resistance to imperialism, its rocks engraved with quotations from Leader Kim Il-Sung, alongside the engravings left by centuries of earlier visitors. No longer focuses of Buddhist pilgrimage or retreats for the literati, and no longer the Japanese Empire's answer to the Alps, the mountains were reconceived in a strictly national and heroic form best captured by the revolutionary opera *The Song of Kumgang-san Mountain*:

Oh, great is our sun,
Brilliant is the name of Marshall Kim Il Sung.
For the regeneration of this beautiful country
He fought for 15 long years through ordeals.

The towering peaks and the crystal-clear streams
Sing the praises of Marshall Kim Il Sung's benevolence . . .
Were it not for our bright sun
How could Kumgang-san be shining today! (n.a., 1974: 22–23.)

The only foreigners to gain access to the mountains were a few honored official visitors, most of them from China, the Soviet Union, or Eastern Europe. North Korea's international flows, such as they were, stretched northward and westward, while connections to the immediate south were utterly severed. Interestingly, among the very few Japanese visitors in the Cold War years were the artists Maruki Iri and Toshi—renowned for their powerful depictions of the atomic bombing of Hiroshima and Nagasaki—who were invited by the North Korean government to a conference it organized in 1956 to mark the 450th anniversary of the death of the artist Sesshū. During the visit, Maruki Toshi reportedly stated that she hoped to include the landscape of Geumgang in the background of the couple's next work in the Hiroshima series, though I can find no sign of the mountains in the finished version of the series.[10]

The Long Path to Reintegration—From the 1980s Onward

The traditional approach to area studies tends to see the force that constitute the area as arising out of the bedrock of shared culture. By contrast, a liquid area studies approach emphasizes how much these flows are products of contemporary economic, political, and social forces. Rather than a stable cultural commonality producing economic and political cooperation, we might argue that it is the contact created by trade, migration, and travel that leads people to seek out, rediscover, and reinterpret cultural and historical commonalities. From this perspective, the opening up of Mt. Geumgang to new flows of people during the first years of the twenty-first century can be seen as exemplifying both the possibilities and perils of reintegration in contemporary Northeast Asia.

Towards the end of the second decade of the twenty-first century, a century after the Japanese annexation of Korea, the people of Northeast Asia are still in the process of gradually recreating the regional flows severed by the Cold War. From the 1980s onward, the collapse of the Soviet Union and the transformation of the People's Republic of China opened the way to a remarkable relinking of places that, though geographically close to one another, had been separated for decades by ideological barriers. The physical infrastructure that sustains these new flows has again been transformed—this time, air routes play a key role alongside highways, railways, and shipping routes.

In the two decades to 2017, economic growth and increasing integration made the Northeast Asian region the world's second largest trading bloc after the European Union (United Nations 2017: vi). Rapid Chinese economic growth both sustained and was sustained by large investment flows within the region. The annual inflow of foreign direct investment to China increased more than fourfold between 2002 and 2007, with a large share of this being generated by intraregional investment from Japan, South Korea, and Taiwan (UNCTAD 2008: esp. 2). Between 2010 and 2014, more than one-third of Japan's foreign direct investment went to other countries of Northeast Asia, particularly China, though China's share in the total has since declined somewhat as costs of production in China have risen (United Nations 2017: 22). The levels of integration vary not just by country but by region—for example, northeastern areas of China close to the Russian border depend very heavily on trade on investment links with Russia, while southeastern areas are more oriented toward distant markets like the United States.

Meanwhile, regional flows of cross-border travel and migration emerged, some retracing older paths of human movement, others carving out new trajectories. Ethnic Koreans—descendants of colonial-period emigrants to Manchuria—migrate in large numbers to South Korea. As of 2015, more than 1.4 million Chinese people were living as first-generation migrants in other parts of Northeast Asia (almost half of them in Japan), where Chinese constituted the largest group of foreign residents (United Nations 2017: 27; Japan Statistics Bureau 2017). Large numbers of South Korean students make study visits to China, and large numbers of Chinese students take courses at Japanese colleges and universities. Young Japanese seek jobs in Dalian and Shanghai; Chinese and Korean visitors contributed to the boom in the number of tourists visiting Japan, which doubled between 2014 and

2017 (JTB 2018). Meanwhile, despite the dangers, North Koreans also make undocumented journeys into China, some crossing back and forth to trade, while others seek more permanent refuge outside their homeland.

As Akaha Tsuneo and Anna Vassilieva (2) point out, "there are signs throughout the region that a major change is afoot. Increasing numbers of ordinary citizens in all Northeast Asian countries are finding it necessary, desirable, and indeed possible to travel to neighboring countries. Some of them decide to settle permanently in the host society, others find temporary employment as migrant workers and still others travel simply as tourists." The new wave of Northeast Asian regional flows has been sustained by globalization and market economic growth, particularly centered on eastern China. But these flows are impeded and complicated by a variety of factors, of which the most important is the still-unresolved fragment of the Cold War that has survived on the Korean Peninsula, obstructing social and economic contact and creating a perpetual mood of insecurity. Ideology continues to determine just when and how spatial flows are severed or reconnected.

After an agreement was signed between the North Korean government and the Hyundai Asan corporation in 2000, Mt. Geumgang became the site of a huge tourism complex, aimed mainly at South Korean visitors. The complex was also used for diplomatic meetings and for reunions of Korean families long-divided by the Cold War, and the new joint venture was seen as a harbinger of a future gradual reintegration of North Korea into the region. Though tight controls were in force to prevent tourists from mingling freely with local people, these new pilgrims to Mt. Geumgang had an opportunity to meet and talk to North Korean guides and hotel workers and to see something of the surrounding countryside, as well as of the glorious scenery of the mountains. Such simple human contacts undoubtedly helped to chip away at the decades of fear and hostility that had separated the two Koreas. Alongside South Koreans who came to Mt. Geumgang as tourists or guides, or to run small shops and restaurants, ethnic Koreans from China were also brought in to work on the project. As Koreans from North and South and from China came together, they rediscovered the ancient shared history contained in the artistic heritage of the mountains.

This re-remembering, however, involved its own form of forgetting. Now the colonial world of the Mt. Kongō Electric Railway, and the Taishō Period Japanese artistic and literary visitors to the mountains was almost entirely forgotten. The forgetting reflected the wider absence of Japan from key aspects of the integration underway in the first years of the twenty-first century. Transfixed by the anger and hostility that followed revelations about the 1970s–1980s kidnappings of Japanese citizens by North Korea, both the Japanese government and Japanese corporations developed a "North Korea allergy" that led to an avoidance of participation in the gradual opening of North Korea to trade and foreign investment. Though a few Japanese visitors did join tours to Mt. Geumgang, their numbers were extremely small. When I visited the mountains in May 2008, I saw hundreds of fellow tourists, including a sprinkling of Europeans, but not a single Japanese person. This absence seemed symbolic of a much greater absence: the reluctance of Japan to take part in the Six Party process on North Korea and the general hesitancy of Japan in playing an active role in key issues of regional political and economic interaction.

The tentative reintegration of Mt. Geumgang into the region, then, highlighted the ambivalent Japanese approach to regional integration. Another important feature of regional integration highlighted by the tourism project was the great economic and social inequalities created by the market economy whose growth sustained regional integration. South Korean tourists and North Korean local people met at Mt. Geumgang on vastly unequal terms. The brightly lit, four-star hotels of the tourism complex inhabit an enclave surrounded by people whose lives are beset by malnutrition and chronic energy shortages. The tourism venture thus epitomized the profound social, economic, and political problems that still remain to be addressed as regional reintegration proceeds. These deep inequalities, together with a dramatic political shift by the South Korean regime of Lee Myong-Bak away from engagement and toward a revival of Cold War policies may have been factors behind the shooting at Mt. Geumgang in July 2008. They were certainly reasons why this tragic incident, instead of being dealt with through collaboration and negotiation, resulted in the closure of the entire project. Despite efforts to attract Chinese tourists, the mountain resort became a ghost town that sprang to life only briefly on rare occasions (most recently in 2015 and again in early 2018), when it was used as a venue for the all-too-brief reunions of families divided between North and South Korea.

The Shape of Things to Come

If we view Northeast Asia today from those mountain peaks where Gweon Geun once looked down on "the chaotic state of heaven and earth," we see a region again in a moment of flux and uncertainty. A plaque that stands in the Mt. Guemgang tourism complex depicts the mountains as the center of the Korean nation, with links extending north and south and even to the disputed island of Dokdo/Takeshima halfway between Korea and Japan. But at the time of writing, the links to the south remain extremely tenuous. Family reunions in the mountains have been resumed, and South Koreans from the Jogye Buddhist order have held a joint prayer meeting with officials of the North Korean Central Committee of the Buddhist Federation at the Singye-Sa temple, rebuilt between 2004 and 2006 with South Korean support (Mu 2009). But the tourist resort itself remains closed. The renewed dialogue between North and South Korea that followed from the April 2018 meeting between South Korean President Moon Jae-In and North Korean leader Kim Jong-Un has reignited visions of a new inter-Korean transport networkthat would link the Peninsula to China's grand schemes for new "silk routes." If realized, these could once again transform the protean shape of the region, but the prospects for the success of the vision are still uncertain.

Mt. Geumgang today is an unstable vortex. The flows that meet there are propelled by ideology and extend not only to Pyongyang and Seoul but also to Washington, DC, and Beijing. Viewing the scene from Mt. Geumgang, we can see that the divisions and fractures that so many US scholars identify in Northeast Asia can only be understood when we recognize the force of the flows that still today weaves America itself into the Northeast Asian region. What the past has shown is that regions are created both through grand political strategies and through the multitude of small connections woven by ordinary people as they travel and communicate across borders. The approach to areas sketched here suggests that there is nothing inherent in Northeast Asian geography, culture, or history that prevents greater integration, but also that the search for integration cannot rely on appeals to timeless common values such as Confucianism or group consciousness. The flows and vortices that make a region are created, sustained, or severed through human effort, whether violent or peaceable. The future of a vortex like Mt. Geumgang builds on millennia of history but is built by a multitude of small human actions that affirm some elements of the past while negating others. And the destiny of such small vortices have the power to influence the future integration or disintegration of the region as a whole.

Notes

An earlier Japanese-language version of this article was published in the journal *Tagengo Tabunka: Jissen to Kenkyu*, volume 2, number 12.

1 On the problems of nationalism in Northeast Asia, see, for example, Rozman 2003; Qi 2005.

2 On artists and travel, see, for example, Sherman E. Lee's (1991a) "Art in Japan 1450–1550" and "Korean Painting of the Early Choson Period" (1991b), both in Levenson 1991: 215–328, 333–36.

3 The practice of sending monks to collect alms in cities like Chemulpo (Incheon) was still continuing in the late nineteenth century; see Campbell 1891: 9. Campbell was the first Westerner to leave a detailed account of a visit to Mt. Geumgang, and his report (despite its inevitable nineteenth-century European biases) is a valuable source of important information on the region.

4 Yi Guk's (1999) account of his journey, *Dongyugi*, is reproduced in Kim Dong-Ju 1999: 54–70.

5 I am indebted to Dane Alston for making this poem known to me.

6 Accounts of travels to Mt. Geumgang by literati included Nam Ho-Un's *Yugeumgansangi* (1485), Seong Jaeweon's *Yugeumgangrok* (1531), Yi Seong-Gu's *Yugeumgangsangi* (1603), Kim Chang-Hyeop's *Dongyugi* (1671) and Seok Beob-Jong's *Yugeumgangrok* (1670).

7 See Kwon 1993: 171–218; on Seosan and Samyeong, see Grayson 2002: 122–23.

8 Kongō is the Japanese pronunciation of the place name Geumgang. The Mt. Kongō Electrical Railway was constructed over an eight-year period between 1923 and 1931; see Kongōsan 1939.

9 Chōsen Government Railways advertisement included in Terry 1928.

10 "Name List of Japanese Supporting North Korean Policies and/or Visiting North Korea (as noted in Radio Pyongyang broadcasts 16 August 1956 to 15 May 1957)." Part 1, *Japan: Relations with North Korea*, ser. 1838, no.1303/11/91. Australian National Archives.

References

Akaha, Tsuneo, and Anna Vassilieva. 2005. "Introduction." In *Crossing National Borders: Human Migration Issues in North East Asia*, edited by Tsuneo Akaha and Anna Vassilieva, 1–7. Tokyo: United Nations University Press.

Alston, Dane. 2008. "Emperor and Emissary: The Hongwu Emperor, Kwŏn Kŭn, and the Poetry of Late Fourteenth Century Diplomacy." *Korean Studies* 32: 104–213.

Andaya, Barbara Watson. 2007. "Oceans Unbounded: Transversing Asia across 'Area Studies.'" *Japan Focus*, 17 April. apjjf.org/-Barbara-Watson-Andaya/2410/article.html.

Anderson, Benedict. 1978. "Studies of the Thai State: The State of Thai Studies." In *The State of Thai Studies*, edited by Eliezer B. Ayal, 193–247. Athens: Ohio University Center for International Studies, Southeast Asia Program.

Bauman, Zygmunt. 1998. *Globalization: The Human Consequences*. New York: Columbia University Press.

Braudel, Fernand. 1995. *A History of Civilizations*. London: Penguin.

Brovko, P. F., and N. I. Fomina. 2008. "The History of Establishment of the National Park Network in Countries of the Asia-Pacific Region." *Geography and Natural Resources* 29, no. 3: 211–25.

Calder, Kent. 2001. "The New Face of Northeast Asia." *Foreign Affairs* 80, no.1: 106–21.

Campbell, C. W. 1891. "Report by Mr. C. W. Campbell of a Journey in North Korea in September and October 1889." In *British Parliamentary Papers: China*. Vol. 2. London: HM Stationary Office.

Curzon, George N. 1894. *Problems of the Far East: Japan-Korea-China*. London: Longmans, Green, and Co.

Department of Railways. 1920. *An Official Guide to East Asia: Chōsen and Manchuria*. Vol. 1. Tokyo: Department of Railways.

Grayson, James Huntley. 2002. *Korea: A Religious History*, rev. ed. London: Curzon.

Gweon Geun. 1936 [15th day of the 9th month, 29th year of Hongwu]. "Geumgangsan." In *Ten Poems on Prescribed Themes*, translated by Dane Alston.

Hamashita Takeshi. 1999. *Higashi Ajia Sekai no Chiiki Nettowâku*. Tokyo: Kokusai Bunka Kōryū Suishin Kyōkai.

Hamashita Takeshi. 2003. "Tribute and Treaties: Maritime Asia and Treaty Port Networks in the Era of Negotiation." In *The Resurgence of East Asia: 500, 150, and 50 Year Perspectives*, edited by Giovanni Arrighi, Takeshi Hamashita, and Mark Selden, 17–50. London: Routledge.

Hye Ch'o. 1985. *The Hye Ch'o Diary: A Memoir of the Pilgrimage to the Five Regions of India*, translated and edited by Han-Sung Yang. Berkeley, CA: Asian Humanities Press.

Japan Statistics Bureau. 2017. *Statistical Handbook of Japan, 2017: Foreign Residents by Nationality*. www.stat.go.jp/data/nenkan/66nenkan/zuhyou/y660210000.xls.

Itō, Takeo. 1988. *Life along the South Manchurian Railway: The Memoirs of Itō Takeo*, translated by Joshua A. Fogel. Armonk, NY: M. E. Sharpe.

Iyotani, Toshio. 2005. "Migration as Method." In *Motion in Place/Place in Motion: Twenty-First Century Migration*, edited by Iyotani Toshio and Ishii Masako. Osaka: Japan Center for Area Studies.

Jackson, Peter. 2003. "Space, Theory, and Hegemony: The Dual Crises of Asian Area Studies and Cultural Studies." *Sojourn: Journal of Social Issues in Southeast Asia* 18, no. 1: 1–41.

Jeong, In-Gap. 1992. *Cheonhwa Jaeil Myeongsan Geumgangsan*. Seoul: Eseupero Mungo.

JTB Tourism Research and Consulting Company. 2018. "Japan-Bound Statistics." www.tourism.jp/en/tourism-database/stats/inbound/#annual.

Kim, Dong-Ju, ed. 1999. *Geumgangsan Yuramgi*, Seoul: Jeontong Munhwa Yeonguhoe.

Kim, Chang-Hyeop. 1671. *Dongyugi*.

Kim, Ok-Seon. [1938] 2007. "Geumgangsan Tamseunggi." In Kim Yeong-Seok et al., *Sin-yeoseong, Gil uie Seoda*, 29–38. Seoul: Homi.

Kim, Soung Chul. 2008. "Multilateral Security and Economic Cooperation in Northeast Asia." *Sejong Seongchek Yeongu* 4, no. 2: 225–97.

Kongōsan Denki Tetsudō Kabushiki Kaisha. 1939. *Kongōsan Denki Tetsudō Kabushiki Kaisha Sanjūnenshi*. Tokyo: Kongōsan Denki Tetsudō Kabushiki Kaisha.

Kwon, Kee-Jong. 1993. "Buddhism Undergoes Hardships: Buddhism in the Chosŏn Dynasty." In *The History and Culture of Buddhism in Korea*, edited by Korean Buddhist Research Institute, 171–218. Seoul: Dongguk University Press.

Lee, Sherman E. 1991a. "Art in Japan 1450–1550." In Levenson *Circa 1492*, 215–329.

Lee, Sherman E. 1991b. "Korean Painting of the Early Choson Period." In Levenson *Circa 1492*, 333–36.

Lefebvre, Henri. 1991. *The Production of Space*, translated by Donald Nicholson-Smith. Oxford: Blackwell Publishing.

Levenson, Jay, ed. 1991. *Circa 1492: Art in the Age of Exploration*. New Haven, CT: Yale University Press.

Mantetsu Keijō Tetsudō Kyoku, ed. 1924. *Chōsen Kongōsan*. Tokyo: Mantetsu Keijō Tetsudō Kyoku.

Ministry of the Environment. 2017. "National Parks of Japan." Japan. www.env.go.jp/en /nature/nps/park/index.html (accessed November 15, 2018).

Morris-Suzuki, Tessa. 2000. "Anti-Area Studies." *Communal/Plural* 8, no. 1: 9–23.

Mu Xuequan. 2009. "Inter-Korean Buddhist Mass Praying for Reunification Held in DPRK," *China View*, October 12. news.xinhuanet.com/english/2009-10/13/content_12225682 .htm

N.a. 1974. *The Song of Kumgang-san Mountain*. Pyongyang: Foreign Languages.

Nam Ho-Un. 1485. *Yugeumgansangi.*

Park, Hyun-Ok. 2005. *Two Dreams, One Bed: Empire, Social Life, and the Origins of the North Korean Revolution in Manchuria*. Durham, NC: Duke University Press.

Qi, Zeng. 2005. "The Clash of Nationalism in Northeast Asia in the Transnational Context." In "Building a Stable Northeast Asia: Views from the Next Generation." Special issue, *Pacific Forum CSIS Issues and Insights* 5, no. 12: 1–51.

Rogers, Larry B. 2004. "Northeast Asia: Cultural Influences on the US Security Strategy." In "Carlisle Papers," *Studies in Intelligence*: 1–14. https://www.globalsecurity.org/military /library/report/2004/ssi_rogers.pdf

Rozman, Gilbert. 2003. *Northeast Asia's Stunted Regionalism: Bilateral Distrust in the Shadow of Globalization*. Cambridge, MA: Cambridge University Press.

Scott, James C. 2009. *The Art of Not Being Governed: An Anarchist History of Upland Southeast Asia*. New Haven, CT: Yale University Press.

Seok, Beob-Jong. 1670. *Yugeumgangrok.*

Seong, Jaeweon. 1531. *Yugeumgangrok.*

Snow, Helen Foster. 1984. *My China Years: A Memoir*. New York: Morrow.

Terry, T. Philip. 1928. *Terry's Guide to the Japanese Empire*. Rev ed. Boston: Houghton Mifflin.

Tokuda, Tomijirō. 1918. *Kongōsan Shashinchō*. Tokyo: Tokuda Shashinkan.

Tsai, Shih-Shan Henry. 1995. *The Eunuchs in the Ming Dynasty*. New York: State University of New York Press.

UNCTAD. 2008. *World Investment Report 2008*. New York and Geneva: United Nations.

United Nations Economic and Social Commission for Asia and the Pacific. 2017. *Unlocking the Potential for East and North-East Asian Regional Cooperation and Integration*. Incheon, South Korea: United Nations.

Van Schendel, Willem. 2002. "Geographies of Knowing, Geographies of Ignorance: Jumping Scale in Southeast Asia." *Environment and Planning D: Society and Space* 20, no 6: 647–68.

Wallerstein, Immanuel. 1998. "The Unintended Consequences of Cold War Area Studies." In *The Cold War and the University*, edited by Noam Chomsky, Laura Nader, Immanuel Wallerstein, Richard C Lewontin, and Richard Ohmann, 195–232. New York: New Press.

Wright, Logan. 2006. Review of *Northeast Asia's Stunted Regionalism*, by Gilbert Rozman. *Yale Journal of International Affairs* 1, no. 2: 172–75.

Wu, Pei-Yi. 1992. "An Ambivalent Pilgrim to T'ai Shan in the Seventeenth Century." In *Pilgrims and Sacred Sites in China*, edited by Susan Naquin and Chün-fang Yü, 65–88. Berkeley: University of California Press.

Yi, Guk. 1999. *Dongyugi*. In *Geumgangsan Yuramgi*, edited by Kim Dong-Ju, 54–70. Seoul: Jeontong Munhwa Yeonguhoe.

Yi, Seong-Gu. 1603. *Yugeumgangsangi*.

Yu, Hong-Jun. 1998. "Geumgangsan ui Yeogsa wa Munhwa Yusan." In *Geumgangsan*, edited by Yu Hong-Jun, 11–38. Seoul: Hakgojae.

The Regime of Separation and the Performativity of Area

Naoki Sakai

The Ambiguity of the Area: Performativity and Fixed Location

In our inquiry into area studies, area may, first of all, appear to be a geographic index, the circumscribed region of territory, community, or social institutions that is coordinated in relation to other geographic indices. One may be tempted to start with the assumption that the area is a location or identifiable spatial index, which is, from the outset, accommodated within the space of a geographic location. It is important to note, however, that the area in the disciplinary formation of area studies refers not only to the determined locality in a geographic configuration; it must also and always designate the coming into being of geographic order without which a spot, a border, or an enclosure cannot be *located*. In other words, the area is an act or performativity corresponding to the grammatical category of verb,

positions 27:1 DOI 10.1215/10679847-7251910

while it is also taken to be a noun. The area is ambiguous precisely in this respect. In order to render a geographic *location* meaningful at all, what must be introduced into an otherwise amorphous material flow or *machinic phylum* is an order of measurement, axes of scale, or a system of a grid.[1] For it is impossible to identify the position of territory, a community, or a sovereignty geographically and geopolitically unless the portion of the earth's surface in question is ordered with respect to its spatial measurability, unless it has been transformed into a space for comparison. To identify an area is to inscribe it within the order of spatial coordinates in relation to other localizable references, thereby rendering a space in which to compare an area externally to other areas. In short, the area signifies not only a determined location in the geographic space ordered by latitude and longitude; it also implies a transformation of space from a smooth one into a striated one, to borrow the vocabulary of Gilles Deleuze and Félix Guattari (1987).

At the same time, an area presents itself as an assembly or congregation of multitudes of things that are in one way or another qualitatively similar to one another, or next to one another to form a neighborhood; an area is recognized as if it were an enclosure internally unified. It follows that the components of an area are expected to share some commonality; they are supposed to be homogeneous. In turn, the commonality of an assembly or congregation is most often represented by the figure of an area. It is well known that in area studies the figure of a national or ethnic culture or language has so often been confused with an area itself, so that it has been very difficult to evade the presumption that, as most typically observable in the early phase of area studies usually referred to as "national character studies," an area designates a geographical extension of common culture or language. Deprived of its projective and mapping performativity, the area thus stands in the juncture of similarity and dissimilarity: an internal homogeneity in which the components are similar to one another and an external heterogeneity in which members of one area are dissimilar to the members of another area.

We must keep in mind that such an economy of territorial homogeneity and heterogeneity is rather a recent invention. One cannot take for granted that the entire surface of the earth has always been apprehended as a space for location and comparative measurement. On the contrary, it is only in the

Age of Discovery that the global order of the earth—what Carl Schmitt (2006) called *nomos*—was brought about globally and further consolidated into the Eurocentric system of international law. The discovery of the New World prompted the first nomos of the earth and thus marked the beginning of what we would later call "the modern international world." It is no accident that the international world came into being almost simultaneously with the technology of modern cartography. As Tongchai Winichakul (1994) has brilliantly illustrated in his *Siam Mapped: A History of the Geo-Body of a Nation*, the process of transforming state sovereignty and the order of geographic space must be repeated each time a new nation is built. In this respect, the internationality of the modern international world cannot be done once and for all; it must be repeated as long as it sustains itself as an institution. Of course, this reminds us that the historicity of internationality must be conceptualized in a manner homologous to that of "the primitive accumulation of capital."[2]

With a view to the history of the modern international world, therefore, we cannot overlook the morphological ambiguity inherent in the term *area*. Just as the very process or operation in which the order of geographic measurement and standard is projected onto a machinic phylum in the notion of "location," which connotes location as a geographic index—a determined spot within the system of geographic coordinates—as well as location as an act of mapping or of comparative determination, the area too of necessity comprises the two aspects of the determination of the putative object of this disciplinary formation called "area studies." An area is a circumscribed geographic region that serves as a frame for knowledge production in a discipline of area studies, but it also connotes a conduct or operation in which a particular geographic expanse is mapped as such. As long as we ignore its second aspect, we will remain unaware of the historicity of the very idea of area itself, of the very apparatus in which an area is postulated, together with a particular conduct of mapping. What I delineate in the following is an inquiry into the historicity of the very conditions under which an area has been routinized as a frame—and also framework—for knowledge production, as well as an apparatus of mapping for such a disciplinary form of knowledge production.

An Epistemology of Race Recognition and the Logic of the Area

Area studies is the specific name for a disciplinary formation: it was a new formation of academic disciplines institutionalized at universities and research organizations in the United States after the Second World War, and, in contrast with the four centuries of the history of the modern international world, area studies is a relatively recent convention largely confined to American higher education, even though it has been increasingly adopted in educational systems outside the United States in the late twentieth and early twenty-first centuries.

A few early area specialists insisted that the United States' hegemony, which area studies was expected to promote and serve, marked a decisive departure from the previous forms of Oriental and African studies closely affiliated with European colonial administrations; but it is undeniable that area studies grew out of what Stuart Hall (1996) called "the discourse of the-West-and-the-Rest," a power arrangement particular to the modern international world. Initially, in the sixteenth and seventeenth centuries, the discourse of the-West-and-the-Rest was intimately associated with the theological gesture of dividing the world, distinguishing the pagan world of *theirs* from the Christendom of *ours*. As European societies evolved from the many conflicts of the Reformation, and as the international system of territorial *national* sovereignty was introduced in the eighteenth and nineteenth centuries, the bifurcation of the world became secularized, so that the direct connotation of religious bigotry became less emphasized. Yet, even today, one cannot overlook the aspect of identity politics in the very notion of the area, for its notion is still couched within the discourse of the-West-and-the-Rest. In this respect, the area in the disciplinary formation of area studies cannot be examined without reference to the division of the world in terms of the West and the Rest.

The word *area* may be substituted for *region, domain, territory, vicinity, section, locale*, or *spot*, depending on the context of a discussion. The relevance of any such substitution to a large extent relies on the semantic and tropic function of the word, so that it may highlight its spatial connotation (*region, territory, spot*) or its tropic working in classification (*domain, section*); it may draw attention to the site of performance (*stage, vicinity, locale*). In the discipline of area studies, however, area acquires a historical specificity, an elucidation of which demands that we pay more attention to the economy of such polysemous substitutions and deliberately distinguish the term from its semantic equivalents.

By equating area to territory, for instance, some distinctive traits of area operating in knowledge production in the discipline of area studies would most likely be ignored and repressed. Such an equation is justified as long as the area is apprehended exclusively in a geographic configuration as an object about which knowledge is produced in area studies. One aspect of the area that is completely disregarded when an area is equated to a territory—and this happens routinely in the practices of area studies—is the very aspect of its *location* not as a nominal but as performativity, of the introduction of a geopolitical division, or the act of mapping the world, that is, the originary conduct by which the world is presented as a stage for comparison. What is necessarily implicated in the area is the identification of an object of knowledge in relation to the agent who knows it and speculates on it, an identification spatial in essence because the introduction of spatial division entails geographic *separation*, according to which the agent of knowledge and the object of knowledge are assigned to two distinct positions. Of course, this *separation* cannot be apprehended without reference to what Jon Solomon refers to as *anthropological difference*.

Yet, it is important to note that this separation is oxymoronic, so to say, or inherently contradictory, since it is not a separation of one term from another on the same plane but rather a separation involving a certain incommensurability. As an anthropological difference, it is difference in the epistemic sense *and* the practical sense at the same time: one sort of humanity is factually and naturally different from another sort, but it is demanded that it should be different from the rest of humanity. It is precisely in this respect that the area is inherently ambiguous, always loaded with the two and incompatible senses of location: location in the sense of determined geographic index on the one hand, and location in the sense of erecting an order of hierarchy among the species of humanity in general on the other.

Accordingly, the ambiguity of location solicits us to return to the topic of what I have called the *regime of separation*, whose workings I detected as underlying the institutionalized practices of area studies. In the inaugural issue of the renewed focus of this journal, with its new name, *positions: asia critique*, I attempted to analyze it with respect to its historical, political, and epistemic ramifications (Sakai 2012: 67–94). Now readers can see why I insist on calling into question the routinized equation of the area to the territory, which in fact has been customarily accepted in the practices of area studies. It is precisely because the regime of separation has been rendered invisible or unproblematic that the area has been treated as if it were synonymous with the territory. As a result, many have taken it to be an already established truism that the area can be substituted for the territory. No doubt, this is partly because the concept of territory has played such an overwhelmingly decisive role in the modern international world, in which the central political entity was neither the monarchical head of a large kingdom nor the theocratic authority of politico-religious governance, but a secularized agency called "the territorial state sovereignty," which is routinely regarded as synonymous with "people," increasingly so after American Independence and the French Revolution. In this respect, the concept of territory was a marker of modernity in international politics and jurisprudence. Very often in modern history, *state ligitimacy* has been defined in terms of the territorial integrity of the state, which supposedly reflects the homogeneity of its resident population.

Not surprisingly, many existing disciplines of area studies—Chinese studies, Soviet studies, Japanese studies and so on—have been institutionalized on the premise that equates the territory of the state sovereignty to the area; as a result of this the area may often appear synonymous with its territory and the population inhabiting it. In contrast, such disciplinary formations of area studies as Latin American studies, Southeast Asian studies, and African studies, are free from such confusion. But the point is not the empirical validity of the concept of territory with regard to area. What is at task is that we must draw attention to the problem of spatial ordering or bordering as implicated in the concept of the area, whose presence we cannot afford to overlook, and thanks to which an object of knowledge is constituted with a system of comparability.

To further elucidate the significance of this aspect of the area, therefore, please allow me to appeal to a classical text, an encounter with which opened my eyes for the first time to the general problem of *separation* or *bordering* more than three decades ago. The text that taught me so much is usually supposed to belong not to such genres as sociology, geography, philosophy, and intellectual history, but rather to literature: it is *The Fire Next Time* by James Baldwin (1963). Though it is lengthy, let me reiterate the portion of it that introduced me to the problem of separation—or bordering and therefore to the general topic of anthropological difference—and in which I could unearth so rich a wealth of ideas and insights that I have repeatedly returned to this piece. In this short book as well as other writings by Baldwin, his insights into the history of modernity are woven together with his incredible perceptiveness and political judiciousness, which overwhelmed me and forced me to call into question my routinized ways of feeling and thinking. As a matter of fact, I had to invent a number of idioms—such as "the schematism of co-figuration" and "the regime of translation" (Sakai 1997) through repeated conversations with this piece of writing, a personal letter addressed to the narrator's nephew:

> There is no reason for you to try to become like white people and there is no basis whatever for their impertinent assumption that *they* must accept *you*. And I mean that very seriously. You must accept them and accept them with love. For these innocent people have no other hope. They are, in effect, still trapped in a history which they do not understand; and until they understand it, they cannot be released from it. They have had to believe for many years, and for innumerable reasons, that black men are inferior to white men. Many of them, indeed, know better, but, as you will discover, people find it very difficult to act on what they know. To act is to be committed, and to be committed is to be in danger. In this case, the danger, in the minds of most white Americans, is the loss of their identity. Try to imagine how you would feel if you woke up one morning to find the sun shining and all the stars aflame. You would be frightened because it is out of the order of nature. Any upheaval in the universe is terrifying because it so profoundly attacks one's sense of one's own reality. Well, the black man has functioned in the white man's world as a fixed star, as an immovable pillar: and as he moves out of his place, heaven and earth are shaken to their foundations. You, don't be afraid. I said that it was intended that you should perish in the ghetto, perish by never being allowed to go behind the white man's definitions, by never being allowed

to spell your proper name. You have, and many of us have, defeated this intention; and, by a terrible law, a terrible paradox, those innocents who believed that your imprisonment made them safe are losing their grasp of reality. But these men are your brothers—your lost, younger brothers. And if the word *integration* means anything, this is what it means: that we, with love, shall force our brothers to see themselves as they are, to cease fleeing from reality and begin to change it. (Baldwin 1963: 9–10)

For this compelling articulation of identity politics to an epistemology of race recognition, Baldwin writes in the form of an epistolary novel a personal experience of racial discrimination in which the structure of modern subjectivity is brilliantly sketched. No doubt, he speaks of race relations and race recognition in terms of the conditions for the possibility of racism in the United States—and in the modern world in general. What he executes in this piece of writing is a sort of transcendental criticism, a criticism of what I would call the dramaturgy of desires. He observes how the very whiteness of the white person is constituted epistemologically but does not neglect the fact that the black is implicated in the racist economy of desires. His astute analysis of race relations never overlooks the emotive aspects of this identity politics of race. This is why it has to be written as a personal letter, from a narrator who is supposed to occupy a specific position in the configurations of the races and social classes in the United States of the 1960s, to an addressee whose racial and social position is equally specified. The voice of persuasion in the letter is not uttered from a neutral or objective stance. At first glance, it may appear that the letter does not expect to be read by readers who would identify themselves with the white at large, but, as a matter of fact, it is the form of epistolary novel that allows it to be addressed to the white or nonblack audience at large. Consequently, white readers are solicited to occupy a peculiar positionality of eavesdropping, from which the very instability of their racial identity can be disclosed. Thus, a nonblack reader is compelled to relive the reality of racial segregation from an entirely different perspective, but it is to be relived in the posture of eavesdropping in which his or her presence is supposedly not seen or recognized.

Let there be no misunderstanding that the letter writer's insights, his assertions, or his proposals can be refuted and discredited on the basis of empirical positivity alone. His insights, assertions, and proposals are of transcendental criticism. As outlined in the putative reader's positionality in the very experience of eavesdropping in this piece of conversation between an uncle and a nephew, Baldwin is primarily speaking of the configuration of viewpoints, the dynamics of desire generated in such a configuration, and the consequent power relations that result from the structured relationships of positionalities in that configuration; this is indeed not independent of the configuration of race relations in American society and in the modern world in general, for race relations are structured by the epistemology of social recognitions. So please allow me to limit my scope to the powerful impacts I have received from Baldwin's writing, but also expand on what I have learned from him in a direction not entirely alien to that of radicalized transcendental criticism or, may I say, a genealogy of the Nietzchean type.

The Fire Next Time consists of two chapters; the first of which is entitled "My Dungeon Shook" and is a letter written by a man called James to his nephew, also a James. The above quote occurs toward the end of this relatively short chapter. The author of the letter addressing his nephew ascribes the presence of racial segregation in the United States to a certain dynamics of identity politics, according to which the two aspects of segregation are simultaneously articulated to one another. The first is the positing of distance or separation, without which the white would not necessarily be distinct from the black. Due to this separation thus posited, the white person is assumed to be dissimilar to the black man. Nevertheless, the separation at issue is not of a factual difference but rather an insistence on distinction: above all else, separation is a matter of desire; it stems from the desire for identity: the white *ought* to be different from the black. For, only when whites are distinguished from blacks can they subsequently identify themselves as white. In other words, they are *putatively* white. Yet, their whiteness is only reflectively or by inference ascertained. The confirmation of their whiteness only comes from blacks. Let us keep in mind that this act of separation consists in the positive positing of the black in the first place; it is impossible to posit the agent who recognizes and speculates on the black as dissimilar, different, or distinguished from him- or herself unless the black is first posited there as an object of such recognition. Accordingly, the second aspect must supplement the first because the distance of the black from the white is not that of one point from another, one position from another, in an already coordinated space of commensurability; the distance of the black from the white is not distance in *location* in the first place.

As I later argue, this separation of the black from the white is, initially, not of *diaphora* (species difference) in the classical logic of generality and particularity, or of *genus* and *species*. Nor are the two positions thus separated in the same space, that is, under a commensurate measurement. In fact, these two are not *comparable*. Yes, it is supposed that a white person can be compared with a black person, and one might conclude that the two are different, dissimilar, and distinct from one another, but such a comparison can be conducted only from the positionality of the third person, who necessarily remains indeterminate as to his or her whiteness or blackness. If the race of this third person is to be determined, one must postulate the positionality of another third person—the fourth person?—from whose viewpoint the third person is compared.

A white person may assert that the black person is different from his or her ego or self. But only the black from whom he or she distinguishes him or herself can be postulated as an object of consciousness. Consciousness being a one-sided relationship—it is always to be conscious of x where, if this x does not exist, consciousness is impossible—the self (me or us) is bound to remain indeterminate. This is to say that the agent who recognizes blackness in the black is colorless, so to speak, and cannot be located in the spectrum of colors. White and black are two particularities in the generality of color; the difference between them is supposedly a species difference (diaphora) between two species in the commonality of the genus color. But, whiteness is not only a species in the spectrum of colors; it must occupy a third position beyond color so as to compare two items within the color spectrum. The epistemology of race recognition in fact betrays this classical economy of particularity and generality; whiteness is outside the spectrum of colors. At least in theory, whiteness as a trait of one who recognizes the black person as black is of no color. In other words, whiteness is a color and colorless at the same time.

It follows that only through the identification of the black can the self of the white can be inferred. Therefore, the black person has served as an anchoring point to maintain the white person's world. Unless the fixity of the black were given, the white person could barely understand where he or she stands or hardly tell who he or she is. In other words, the white person would be lost if his or her identity were not supported by the stable identity of the black.

Immediately here one recognizes the famous *aporia* of transcendental illusion. But, if a white person tries to distinguish himself from a black person, the agent who postulates and recognizes a black person as such cannot be identified empirically, whereas a human figure who is *supposedly* black can be postulated as an object of knowledge. The empirical knowledge of a *putatively* white person by the very agent of such cognition is impossible since it is outside our possibility of experience. It is only through inference or guessing that a white person is assured of his or her whiteness. Only in reference to the attribute of blackness of the black can the white ever postulate him or herself as white. In this respect, Baldwin's observation is precise and penetrating, without meddling with philosophical vocabulary. Let me quote the essential dynamics of race relations once again: "the black man has functioned in the white man's world as a fixed star, as an immovable pillar." In short, only as long as the black person is indisputably black, can the white person be sure that he or she is white. As soon as the blackness of the black person is called into question, the world of the white person crumbles.

In this dramaturgy of racial identities, it is essential to point out the structural asymmetry in the allocation of white and black positionalities; from the outset, whiteness is attributed to the agent or subject of recognition while blackness is to the object of recognition. Who is to know and who is to be known are clearly delineated. For the time being, let us not call into question this structural asymmetry; let us wait until we "act," are "committed," and inevitably are "in danger" so as to frankly propose how to change this history in which we are trapped.

In this dramaturgy, the whiteness of the whites is imaginary in two senses. First, blackness is not an empirically observable property, but a symbolic marker of a person's positionality in the configuration of social interactions. Accordingly, projective imagination is always at work in the determination of who is registered as white and who as black. We already are aware of an abundance of cases in which race recognition has little to do with physiology, physical anthropology, or biology at large. It is widely acknowledged that scientific racism is bankrupt, in spite of its anachronistic popularity. Not only does the category of race have almost nothing to do with the color of a person's skin but also, and more generally, it has little to do with physical features of a person's body either. It is in this sense that the category of race is essentially imaginary, but, by saying this, of course, I do not imply at all that race is illusionary.

Second, it is imaginary in the very sense that the ego or self is an image. What is known as "the order of the imaginary" in psychoanalysis is involved in this determination of the self or *ego*. More broadly, what is intimated in Baldwin's historical assessment of American race relations are the questions surrounding the concept of negativity in the general philosophical discussion of modern subjectivity.

Even though the topic is nothing but race relationship between the black and the white and its history in the United States, it is not hard to detect the implications of this personal diagnosis of the history of American racism in the context of our discussion on area studies. Undoubtedly, regardless of whether or not area specialists are ready to understand this history, the history of the modern international world cannot be cleansed of what we generally refer to as racism.

It goes without saying that what I find insinuated in Baldwin's insights is a question, which may well guide our inquiry into the area in area studies. The question is, can we not see the separation of the West and the Rest in the same vein, in an analogy to the epistemology of race recognition in the United States?

Depending on how one participates in endeavors to interfere with and transform a social formation generally referred to as racism, one may as well take up different viewpoints or focus on particular aspects of the issue: one may be concerned with racial discrimination in a national context, the context of the global history of colonialism, or that of the nation form inherent in racism. Yet, one cannot disregard the general thesis that the formation of area studies, the epistemology of race recognition, the identity politics of races, the international system of the territorial national sovereignty, and so forth, are mutually implicated in one another; often we call this network of entanglements "modernity."

So far in this chapter I have taken into consideration the three major registers of modernity. First, the ambiguity of the area—two senses of location: the performativity of the area and the localized index in a striated geographic space—in the disciplinary formation of area studies; second, the epistemology of race recognition in the dramaturgy of race desires; and third, the bifurcation of the West and the Rest in the discourse of the-West-and-the-Rest in the modern international world. These three registers are mutually affiliated yet can be apprehended independently without reference to the others. Accordingly, I have never claimed either that they are connected in terms of causality or the three aspects of the same substance called modernity, or that they can be treated as independent variables.

Above all else, thanks to Baldwin's insights, I have appreciated what the conduct of the area accomplishes; we have also learned that we have not understood the very history of area studies in which many of us, as area specialists, have been trapped. Boldly and decisively, he articulated the political dynamics of social recognition to the dramaturgy of desires. I want to follow this passage of "acting, being committed, and being in danger" outlined by James Baldwin with regard to the practices of area studies. Then, how should we construe this tapestry of entanglement in which all three registers are rather loosely interwoven? In other words, how should we comprehend this planetary space-temporal configuration of the interdependencies of modernity?

The steps I want to take now are only preliminary. The task of challenging and interfering with racist epistemology in the modern international world cannot be accomplished in one sweep. So allow me to take a gradualist approach to the task at hand. In what follows, you will see my preparatory work on the performativity of the area, since I believe this preparation is necessary for identifying how racism is implicated in the disciplinary formation of our knowledge production.

Similarity and the Classical Logic of Individual, Species, and Genus

It is obvious by now, after a discussion of the epistemology of race recognition and the conduct of the area in area studies, that, above all else, the area is an apparatus in terms of which the perceived as well as the imagined things are classified and related to one another. It is often claimed that an area can be postulated because things found in it are similar to one another, while they are dissimilar to things that belong to the outside of that area. Notwithstanding a great number of dissimilarities, for instance, there are a number of aspects in which they are similar to one another. Thus, the judgment of similarity works powerfully, often justifying on some empirical basis the assembling and classifying manifolds of empirical data into such unities as culture, ethnicity, race, and nation. First of all, therefore, it is necessary to examine the concept of the area for its empirical validity, which operates to discern similarity and dissimilarity for the trope of unification. The thread of Ariadne, for the time being, is similarity.

Similarity is equivocally connected with comparison. Similarity is necessarily concerned either with a judgment issued after the act of comparison or with anticipation that calls for a judgment of comparison. It is given in the form of an assertion or prediction that one item can be judged to have a certain relationship—similarity or dissimilarity—with a different item or items as a result of comparison. In discussing similarity, therefore, we cannot evade a preliminary inquiry into the notion of comparison, on which the alternative judgment of being either similar or dissimilar is premised.

Similarity can be predicated on a wide spectrum of objects in many different disciplines of knowledge. However, I am by no means able to discuss the problems of similarity in general without delimiting the scope of inquiry. Thus, in this inquiry into the area in area studies, the first step is to demarcate the scope of my inquiry to the problem of similarity in the humanities and social sciences that participate in area studies.

In the first place, let me single out the two moments in the act of comparison in area studies; these two moments—we may characterize one as logical and the other as political—can always be discerned when we are engaged in a judgment about similarity, one that plays a decisive role in the procedure of comparison in area studies.

The first is the postulation of the class of *genus* among compared items, of some common element in which items are compared. Comparison is performed between or among unified objects, preliminarily identified as two species, while at the same time comparison is constitutive of the logical dimension of genus where *species difference* (*diaphora*) is discovered, measured, or assessed. Attributed to the class of species are particular cultures, languages, economic systems, political ideologies, literatures, and so forth; subsumed under each of these is postulated as an indivisible unified entity, that is, as an individual. Thus, this culture, language, or literature is compared as that culture, language, or literature, as an individual, and the dimension of comparison is nothing but a species of cultures, languages, or literatures. When we submit even such items as culture and political ideology to comparison, they are each regarded as a unity or individual. Thus, we compare the English language with Chinese, for instance, because these two particular languages belong to the category of languages—more specifically *langues* in the vocabulary of Ferdinand de Saussure—as a species. As long as English is assumed to be a systematicity, it can be regarded as an individual,[3] a unified indivisible entity, but we are far from certain that the item to be compared—English or Chinese in this case—exists at the level of the immediately empirical in the way that classical logic has sometimes attributed the property of empirical existence to the concept of the individual. As one of many particular languages, English is an individual of the species of language. However it must be added that a particular language should be able to be conceived of as an indivisible unity precisely because we want to regard it as an individual. The individual is, after all, an *individuum*, an entity that cannot be further divided. Very often comparison is conducted on the presumption that this basic operation of logical and formal reason is still valid and sustainable, and that the individuality of a particular language is indisputable. The discipline of comparative literature, for instance, had long been legitimated on the basis of this logical economy of individual and species, of individual national literatures and the general literature with which universal humanity—which used to be no other than European humanity—was endowed. It goes without saying that our faith in this logical economy has been challenged many times; today it is commonly agreed that it is a matter of suspect intellectual taste to confess one's faith in this logical economy.

In reference to this topic of similarity, the conceptual economy of individual, species, and genus of classical logic seems to me to be vastly inadequate to comprehend the politics of similarity and difference, particularly in view of Ludwig Wittgenstein's (1968: 31–35) argument about family resemblances. The differential perception of similarity and dissimilarity is not organized according to the classification of genus and species, of generality and particularity. How we attribute a racial feature to a person in an every-day social encounter, for instance, is not primarily dictated by the classical logic imbedded in the system of classification in natural history, zoology, or physical anthropology. On the contrary, the belief that humankind can be coherently and systematically classified into subclasses of races rather justi-fies the normative attribution of racial identity to a person. What underlies this is the very belief that similarities and dissimilarities are already orga-nized by the logical economy of individual, species, and genus in nature. What is important to note in apprehending the differential perception of similarity and dissimilarity is that the conceptual economy of individual, species, and genus constitutes an apparatus of the normative attribution appealed to in organizing the diversity of perceived manifolds. Let us return to the famous no. 66 of *Philosophical Investigations*:

> Consider for example the proceedings that we call "games." I mean board-games, card-games, ball-games, Olympic games, and so on. What is com-mon to them all? Don't say: "These *must* be something common, or they would not be called 'games'"—but *look and see* whether there is anything common to all.—For if you look at them you will not see something that is common to *all*, but similarities, relationships, and a whole series of them at that. To repeat: don't think, but look!—Look for example at board-games, with their multifarious relationships. Now pass to card-games; here you find many correspondences with the first group, but many common features drop out, and others appear. When we pass next to ball-games, much that is common is retained, but much is lost.—Are they all 'amusing'? Compare chess with noughts and crosses. Or is there always winning and losing, or competition between players? Think of patience. In ball games there is winning and losing; but when a child throws his ball at the wall and catches it again, this feature has disap-peared. Look at the parts played by skill and luck; and at the difference between skill in chess and skill in tennis. Think now of games like ring-a-ring-a-roses; here is the element of amusement, but how many other characteristic features have disappeared! And we can go through the many, many other groups of games in the same way; can see how simi-larities crop up and disappear.

And the result of this examination is: we see a complicated network of similarities overlapping and criss-crossing: sometimes overall similarities, sometimes similarities of detail. (Wittgenstein 1968: 31–32).

So as to find similarities among various games—board games, card games, Olympic games, and so on—do we have to apply the classificatory scheme of species and genus to them? For the species of card games or Olympic games to be included in the *genus* of games, do they have to share something in common? For instance, as we have seen with regard to Baldwin's insights into the epistemology of race recognition, is the classificatory scheme of the classical logic already operative in another organization of similarities and dissimilarities in the classification of human beings known as racism? For racism is one of the best examples of classificatory logic of individual, species, and genus. It is not difficult to see a procedure of classification similar to that of games be applied to races as subsets of humankind. Must differences in physiological or behavioral appearances among people be classified according to the logical economy of species and genus, precisely since every one of these different people shares humanity in common?

Although normally or normatively, and in the world in which this normalcy has been almost universally accepted, the classification based on genus and species is not distinguished from descriptive or constative claims about racial identity, people do not receive the classification of races in the immediate perception of persons they encounter; perceived similarity and dissimilarity are not given according to the order systematized in terms of individual, genus, and species. Instead, the operation of racial classification, without which it would be simply impossible to attribute a racial identity to a person, is a retrospective inference on our primordial exposure to the world. The recognition of similarity comes before the identification of a particularity in terms of the economy of genus and species, of generality and particularity. Therefore, in the Wittgensteinian manner of speech, we must *see* but not *think*. It seems to me that in the first instance, racial classification is a matter of thinking—inference—rather than seeing. In other words, the assumption of comparability must be applied before comparison is conducted according to the classificatory scheme of the classical logic in the space of comparison.

In exploring what is entailed in the politics of similarity, therefore, I want to discern a stage prior to the conceptual articulation in terms of genus and species, one where the scheme of classical logic is only one of the alternative strategies. Just as I discussed with regard to location, there must be a stage prior to the identification of a local spot in the striated space of comparability, a stage in which the very order of comparison must be instituted. In short, I want to acknowledge the act of introducing the very terms of comparison for genealogical analysis in the topic of similarity. Of course, this leads to the other moment in the act of comparison.

This second moment is the occasion or locale where we are obliged to compare. Comparison takes place because the determination of species difference is needed. But, so as to discern certain political complexity involved in the notion of the determination of species difference, let me ponder what is often referred to as *cultural difference*, over two distinct approaches to cultural difference, for I cannot overlook a certain confusionism in this idiom.

In the presence of a person who appears to be speaking, I can neither understand what he or she wants nor what he or she is meaning to do. Consequently I am at a loss. At such a locale, naturally and in due course, some explanation as to why I or we are at a loss is demanded: reasoning may well provide a schematic explanation about some generalized experience of incommensurability, and about the situation marked by cultural difference. Supposedly what this person speaks is the Chinese language, whereas what I speak is English. Both belong to the general class of languages, but we cannot make ourselves understood to one another because the Chinese language is *different* from English.

Let me pause here momentarily since I do not think that the difference at stake in this instance can in due course be subsumed under the concept of species difference. What is at stake here is *some* difference, but it is not reduced to the conception of difference between Chinese as a particular language and English as another particular language, of difference that conforms to the economy of individual, species, and genus. When I do not understand the Chinese language, how can I possibly judge that it is an individual language, as distinct from the English language, which is subsumed under the general class of languages?

It is worth emphasizing the fact that it is believed that the determination of the species difference is offered as a solution to the initial problem of our being at a loss, in response to the perplexity we come across in such a locale. We, as a collectivity who speak the English language, are different from a different collectivity who speak the Chinese language. Thus species difference exists between the species of Chinese speakers and that of English speakers. This species difference is grasped as cultural difference because of which I, as one of us, cannot understand my interlocutor as one of them. Presumably language difference is a kind of species difference whose purport is to offer a clue about why we are at a loss in a locale of incomprehension, perplexity or helplessness. As I have argued elsewhere, however, the determination of species difference only leads us to a confusionism (Sakai 1997: 1–17). And it is important to note that confusionism of this kind is almost always present in the term cultural difference.

Thus, the language difference is supposed to cause a situation where we need to know why we are at a loss with one another. Language difference, that is already preliminarily determined as species difference, is a serious matter when we do not understand one another, or when we cannot be confident in our directives for the immediate future; it is usually assumed that our sociability is grounded in some primordial communality, on our capacity to be immediately and instantaneously in a common sharing. Language difference is therefore understood to cause our inability to share, to result in the absence of this communality. Normally—we presume normalcy consists of this unwarranted assumption that people do understand each other unless disrupted by some abnormal obstacle—we do not express our doubt about either the comprehensibility of our expressive behaviors or our ability to apprehend others' actions and expressions. The need for comparison occurs only on the occasions where we are forced to become aware of *dissimilar* people, *different* beings whom we are in the presence of. The encounter with cultural difference or discontinuity is interpreted as an encounter with one type or another of species difference.

The term *difference* becomes marked precisely because of this experience of *nonsense, being at a loss*, or *being unable to make sense of the occasion*, in short, of *being deprived of the world*. The determination of species difference becomes something urgent and even desperately important, precisely because we are in the presence of others in *discontinuity*. Most often we talk of this encounter with discontinuity in terms of the foreign, but it is significant that, initially, the foreign does not connote the outside or the external in a strictly spatial sense. Instead it is an *outside* of a given and familiar register, an outside indicative of the nonsensical that evades the spatial alternative of either inside or outside. Initially nonsense is given to us, but it is not given to us as something *localizable*. It is an outside but it cannot be accommodated in the spatial register of our world. It is important to note that the foreign acquires the status of the external or the outsider only through the assumption that we are inside the common element—language, culture, or nationality— represented as an *inside*, spatially delimited by borders. Only when the being-in-common is represented by a spatial figure of enclosure does the foreign gain the trope of the external-to or the outside-of the domain or domesticity or familiarity.[4] And discontinuity is something fundamentally incommensurate with the logic of location in spatial economy. For, at this stage, discontinuity cannot be represented as a relationship within a measured and striated space of comparison. And most importantly it is imperative to keep in mind that it is not because some person or people are *dissimilar* or *different*—in the sense of species difference—from me or us that we are at a loss. On the contrary, it is because we are at a loss or unable to make sense in the first place that we attempt to determine this encounter with difference within the logical economy of individual, species, and genus.

At stake in this presentation is whether or not we are allowed to conceptualize cultural difference in the sense of nonsense or being at a loss in terms of species difference, in terms of the difference categorized in the economy of individual, species, and genus within classical logic. Are we to overlook the very difference between the difference in the sense of nonsense or being at a loss and the species difference already regulated by the logical economy of species and genus?

Allow me to consider another situation where we need to know how we are different from one another, why certain people are not subjugated to the imperatives or commands we normally obey or yield to, or why some of us are free from a set of proscriptions and others are not. Thus we compare ourselves to find where we are situated vis-à-vis one another in a practical sense, along the prescriptive direction of what we must do. Comparison is indispensable precisely because we want to know how we are related to one another, who should lead among us, who should follow, who should work for whom among us, and so on. It is through the act of comparison that we comprehend the configuration of our subject positions in which we apprehend who we are in terms of the socially determinate relations: gender, commodity exchange, race, social class, nationality, civilization, kinship, religion, professional qualifications, pedagogic hierarchy, gift giving, cultural capital, and so forth. When we cannot locate ourselves in the configuration of subject positions, we are also at a loss or do not know how to act in accordance with others in a situation. On such occasions, we sense that something is *different*, and that we ought to behave in such a different manner from the ways in which we are normally and normatively expected to. Yet, this sense of *difference* cannot be simply reduced to difference in the opposition of the same and the different, of the homologous and the heteronomous. Indeed, this is also of crucial importance to apprehend how we understand two different approaches to cultural difference.

In the context of our discussion of cultural difference, I would like to introduce this conceptual ambiguity of area into my understanding of the locale of comparison, of a place where we are articulated to one another in ways that elsewhere I have called "heterolingual" (see Sakai 1997: 1–17).

Transnationality and Internationality

Following this preliminary guideline, let me note that nationality has been one of the most predominant topics, under whose name, if not explicitly, the process of comparison in knowledge production has been addressed in the humanities. Normally nationality signifies some communal premise that guarantees a sympathetic communion among the members of the national community. It is often the case that the term *nationality* suggests the existence of some common ground, thanks to which "we," members of the nation, are supposed to be similar to one another. Yet, the assumed similarity among the members of the nation, of national culture, or of the national society of sympathy cannot manifest without highlighting their dissimilarity to the other nation, other ethnicity, or other culture. As a matter of fact, the homogeneity implied by nationality is most often an inverted image of a nation's relationship to other nations or to the reified presence of the foreign.

In discussing the topics of nationality, therefore, transnationality and internationality must be rigorously distinguished from one another. It goes without saying that neither transnationality nor internationality is intelligible without reference to the operation of comparison. Nationality is essentially a figure, schema, design, or image, resultant from the determination of either transnational or international species difference. What is at issue in transnationality as well as internationality is to compare, distinguish, divide, and draw a border.

In this context, it is worth noting the increasing significance of the problematic of "bordering" in knowledge production today (Mezzadra and Neilson 2008a).[5] This problematic has to be specifically marked as not being one of border but rather of bordering, because what is at issue is considerably more than the old problem of boundary, discrimination, and classification. At the same time that it recognizes the presence of borders, discriminatory regimes, and the paradigms of classification, this problematic sheds light on the processes of drawing a border, of specifying the element of comparison, of instituting the terms of distinction in discrimination, and of introducing a *continuous* space of the social,[6] against the horizon of which a divide is inscribed. I find it particularly significant that Sandro Mezzadra and Brett Neilson first introduced this term *bordering* in what they termed "the multiplication of labor" (Mezzadra and Neilson, 2013).

Through the concept of the multiplication of labor, Mezzadra and Neilson challenge the conventional categorization of labor and the familiar notion of the international division of labor. They question "the orthodoxy that categorizes the global spectrum of labor according to international divisions or stable configurations such as the three worlds model or those elaborated around binary such as center/periphery or North/South" (Mezzadra and Neilson 2008b). What has to be taken into consideration is the dynamic and constantly transforming relationship between labor and power.

The world is not becoming borderless today. On the contrary, borders are constantly redrawn and multiplied; the boundaries of national territories, ethnic cultures, and civilizations are not the only dominant ones; many others are being newly inscribed. Hence, the analytic of bordering requires a simultaneous examination of both the presence of border and its drawing or inscription.

Within the scope of this argument, I want to draw attention to the problematic of bordering so as to elucidate the differentiation of transnationality from nationality. Most importantly I want to reverse the order of apprehension in which transnationality is comprehended on the basis of nationality, on the presumption that nationality is primary while transnationality is somewhat secondary or derivative. This widely accepted pattern of reasoning derives from the thought-habit characteristic of modernity,[7] according to which the adjectival *transnational* is attributed to an incident or situation uncontainable within one nationality: the transnational indicates something that cannot be restricted within the schema or figure of nationality.

What must be highlighted, first of all, is the implicit presumption that underlies the concept of nationality: it cannot make sense unless it is postulated against the horizon of internationality. And we must keep in mind that internationality is utterly incomprehensible unless we take for granted a manner of comparing and categorizing the units—the unities of the nation-states, each of which is represented by state sovereignty—of the world that is to be compared. Essentially, internationality is a historically particular regime of comparison that operates in terms of the economy of individual, species, and genus.

For the very reason of the politics of comparison itself, *nationality* does not make sense unless in conjunction with internationality. Nationality becomes conceivable only when the scene of juxtaposing nations is institutionalized within some scheme of state sovereignty, traditionally designated as territorial state sovereignty. Of course, it is in respect to this affinity of nationality and internationality in our vision of the modern world that a disclaimer must be issued once again: just as the sense of difference concerning the experience of being at a loss must not be equated to species difference, so transnationality must not be confused with internationality. In due course, to distinguish transnationality from internationality is the central issue in the politics of comparison. In order to assert the priority of transnationality over nationality, therefore, our first move is to delineate the semantics of *transnationality* as distinct from *internationality*.

One of the distinguishing characteristics of the modern world can be found in its internationality; the modernity of the modern world has manifested in the formation of the international world. Today, unfortunately, transnationality is all too often understood within the schema of the international world. Here, *schema* means a certain image or figure in terms of which our sense of nationality is apprehended. The schema of nationality is embedded within the larger schema of internationality, and the very relationship between the national and international schemata is understood to be one of species and genus. But it is important to note that in some regions such as East Asia the international world did not prevail until the late nineteenth century. This was also the case in Africa, the Middle East, Southeast Asia, and the Pacific (granted I am using very problematic names for geopolitical regions on the globe). I suspect that this was the case in Northern and Eastern Europe as well, even though the international world had supposedly been established in Western Europe long before.

In East Asia the international world was entirely new; it took more than a century before East Asian states surrendered the old tribute system and yielded to the new interstate diplomacy as dictated by international law. So, it was true of many other regions of the globe that the international world signaled the arrival of colonial modernity. And it was in the very process of introducing the international world that the bifurcation of the West and the Rest began to serve as the framework in which the world's colonial hierarchy was globally actualized and institutionalized.

Of course, the international world is not a phenomenon exclusive to the twentieth century. The dividing of the world into two contrasting areas, the West and the Rest, has been a widely accepted institutionalized practice in academia for a few centuries (Hall 1996). Some argue that this dichotomy may be traced back to the seventeenth century, when the system of international law was inaugurated with the Treaty of Westphalia in 1648 (Schmitt [1950] 2006). This peace treaty, subsequent to the Thirty Years' War, preliminarily established the division of two geopolitical regions. The first would subsequently be called "the international world," in which four principles were to be observed through the reciprocal respect of the diplomatic protocol called "treaty" among the sovereign states of the international world: (1) the territorial sovereignty of the national state and its self-determination,[8] (2) legal equality among national states, (3) the reign of international laws among the states, and (4) the nonintervention of one state in the domestic affairs of another.

The second of these regions was a geopolitical one excluded from the first, inside of which these four principles, including the reign of international laws, had no binding force. The first area where the system of international law was observed called Europe; it would later be also called the West, while the second would be excluded from the international world, and would become literally "the Rest of the world," with its states and inhabitants remaining in an eternal state of exception. In other words, subjected to the arbitrary uses of colonial violence.

Although the area in area studies was initially conceived of independently of this bifurcation of the world, the West and the Rest, it was accommodated in the discourse of it in such a way that the performativity of the area reintroduced and endorsed this bifurcation. Consequently, the area as a determined location can be found only in the Rest, thereby confirming the positionality of the West reflectively in reference to the Rest. Thus the performativity of the area imitates and replicates the dramaturgy of race desire as outlined by Baldwin: just as "the black man has functioned in the white man's world as a fixed star, as an immovable pillar," so the Rest has served in the self-recognition of the West as a fixed star, as the point of anchorage, without reference to which the West cannot be the West.

In East Asia, Japan's colonization of Korea, for instance, was sanctioned according to the protocols of the international world. The Japanese government followed the terms of international law in its colonizing maneuver so as to secure international recognition for its legitimacy. Many parts of the globe were also colonized according to the schema of the international world. By the beginning of the twentieth century, the majority of the second area was transformed into colonies that belonged to a few super powers. Yet, this pseudo-geographic designation of the West—pseudo-geographic because in the final analysis, as we have demonstrated above, the West is not a geographic determinant—gained currency toward the beginning of the twentieth century, gradually replacing the index Europe. At that time the international world expanded to cover the entire surface of the earth as a result of three developments: colonial competition among the imperialist states; the emergence of the United States and Japan as modern imperial powers; and most importantly, the increasingly widespread anticolonial struggles for national self-determination. In this historical characterization of the West, two paramount factors distinguish it from the Rest of the world: the legacy of colonialisms, on the one hand, and the comparative operation based on the logical economy of individual, species, and genus, on the other.

For a colony to gain independence, the colonized had to establish their own national sovereignty and gain recognition from other sovereign states. In other words, the process of decolonization for a colonized nation meant entering the ranks of nation-states in the international world. As the number of nations recognized in the international world increased, the presumptions of nationality and internationality were accepted as natural givens. As the schematic nature of the international world was somewhat forgotten, both nationality and internationality were dehistoricized, as though the institutions that symbolically marked the border of the national community— national territory, national language, national culture, national history, and so forth—had been both naturally inherited and existent for many centuries. Consequently it is no surprise that the vast majority of comparative studies in the humanities and social sciences—comparative literature, comparative law, comparative sociology, and so-called area studies—fall into the general genre of comparative nationality today.

It is at this juncture that the concept of transnationality must be reinvigorated. It must be rejuvenated in order both to undermine the apparent naturalness of nationality and internationality and to disclose the very historicity of our presumptions about nationality, national community, national language, national culture, and ethnicity, which more often than not are associated with "the feeling of nationality." Here, the classical notion of nationality in British Liberalism is of decisive importance to historicize the schema of the international world.

By nationality British thinkers of the nineteenth century meant a new form of communal solidarity that united a portion of humanity to the exclusion of all others. This united portion of humans is called "nation," and is consolidated through shared sympathies. These sympathies are exclusively theirs and cannot be shared by foreigners; they make this segment of humanity cooperate with each other more willingly than with other people, desire to be governed by the government of themselves or of their representatives. John Stuart Mill ([1861] 1972: 391), for instance, called this common sympathy the feeling of nationality; it may be generated by various causes such as the identity of race and descent, language, religion, or shared geographic residence. But the most important factor is the identity of political antecedents; the possession of a national history; communally shared recollections, collective pride and humiliation, pleasure and regret. Through national history this new community could believe that every one of its members was connected with the same incidents in a common past.

A disclaimer must be issued here. As a guiding outline or trope, I introduced the modern binary of the West and the Rest of the world, yet it is imperative to keep in mind that it is no more than a trope, a figurative projection. This binary gives us some synoptic vision of the modern world, but it can hardly sustain coherence in many concrete historical contexts. The task of modernization was equally present in the geographic areas and peoples that can be included in Europe or the West. The very boundary between the West and the Rest is so arbitrarily drawn that there are innumerable cases where certain regions of today's European Union may well be located outside the West. Some social strata of the United States, which have enjoyed the reputation of being the hub of the West's modernity since the end of the Second World War, for instance, manifest the same characteristic of premodernity as is typically attributed to the Rest. Certain aspects of everyday life in some regions in East Asia, for example, are indisputably more "Western" than in some sectors of the "white" population in the United States.

Dependent on the context of comparison, the very distinction between the West and the Rest shifts constantly and articulates differently. Here, it is important to remind ourselves of the theoretical significance of "the multiplication of labor" introduced by Mezzadra and Neilson. In some instances, the bifurcation of the West and the Rest is solidly instituted. In others, it is arbitrary and contingent. It is almost impossible to find a lucid coherence to connect many different manifestations of this dichotomy. Both the West and the Rest are undoubtedly historical and mythical constructs. However, although I would never claim that the West is unreal or illusionary, under every circumstance, I will refrain from talking about it as a kind of transhistorical substance or coherent analytical category.

Translation and Bordering

As is obvious by now, the problematic guiding my inquiry in this chapter is quite different from the nationalist concern shared by many intellectuals in the twentieth and twenty-first centuries. Instead it is committed to the problem of how to emancipate our imagination from the regime of the nation-state, not through negating this regime itself but rather by problematizing the methodological nationalisms that permeate knowledge production in the humanities, in particular in the academic disciplines of area studies, and thereby to present an alternative apprehension of the transnational community. In suspending the nationalist conviction, I refuse to view nationality as a given; instead I reverse the order of priority while never underestimating the nationalist rhetoric mobilized in our struggles with colonial modernity. Simply put, my starting point is that nationality is a restricted and distorted derivative of transnationality. In other words, nationality is always secondary to transnationality. And my guiding question is how the transnational, the primary modality of sociality, is delimited, regulated, and restricted by the rules of the international world. It is in this context that I must confront the issue of *bordering*. In order to problematize the priority of nationality and the international world, the tropics of the border must first be studied.

It goes without saying that the border cannot exist naturally; physical markers such as rivers, mountain ranges, walls, and even lines on the ground become borders only when made to represent a certain pattern of social action. In this respect, a border is always man-made and assumes human sociality. Only when people react to one another does a border come into being. Even if it separates, discriminates, or distances one group from another, people must be in some kind of social relation for that border to serve as a marker or representation of separation, discrimination, or distance. A border is a trope serving to paradoxically and irrevocably represent primordial sociality. Only where people agree to *border* can we talk about a border as an institution. Thus, *bordering* always precedes the border. To apprehend the border is to study how it is inscribed, erased, redrawn or reproduced.

Prior to this bordering, it is impossible to conceptualize the national border. Thus, the national territory is indeterminate prior to bordering. Similarly, it is impossible to determine a national language prior to it. What we conventionally call "language" is not a unity; it is a system but never a systematicity; it does not consist of a finite set of fixed rules; even a grammatical mistake makes sense and becomes an effective expression in language. In its every use, or in what de Saussure called *language* in contradistinction to *langue*, it is modified and reproduced. In short, language is manifold. Therefore, it is impossible to talk about language as if it were a primordial or an indivisible unity; it is never given as an individual. So it follows that without reference to bordering, we cannot comprehend how the individuality of a particular national language, the very indivisible unity of a language presumed in the figure of national or ethnic language, came to prevail. In other words, it is not about the language but rather at the level of its *representation* that its identity and unity are attributable. In due course the operation of comparison, by means of which species difference between languages is postulated, measured, and judged, is impossible unless the individuality of one particular language and another, to be compared at the level of representation, is presumed (see Mezzadra and Sakai 2014: 9–29).

Only at the level of representation can one language be compared with another as an instance of one particularity, as opposed to another within the economy of individual, species, and genus. When we are at a loss, when we come across the situation of incommensurability, when we face nonsense *before translation*, we are in the presence of cultural difference that does not conform to the logical economy of internationality.

So what corresponds to this bordering as far as language is concerned? Of course, it is *translation*. What I want to emphasize here is that in the context of cultural difference, translation comes prior to the determination of language unities that supposedly translation is understood to bridge. Against a commonsensical prejudgment, therefore, it must be insisted that there is translation before the postulation of a national or ethnic language. Translation must occur before the attribution of identity to language, just as there is transnationality before nationality. In short, as far as the representation of languages is concerned, translation comes before the determination of species difference. Only after translation are individual languages available for comparison. In other words, it is in translation that we are able to talk about similarity and dissimilarity of languages. And let us not forget that there is no reason why similarity and dissimilarity thus encountered must necessarily be restricted by the economy of individual, species, and genus.

At this juncture, we can see one reason why it is necessary to touch on the act of bordering, prior to a focus on comparison. Of course, the process of comparison cannot be initiated unless the items to be compared are postulated as comparable. In other words, in the occasion of *discontinuity* in which one is at a loss, unable to comprehend what is going on, and faces nonsense, one cannot even start to compare. Yet, it is this occasion of discontinuity that demands translation. In this particular context of our discussion, what opens the place of comparison is nothing other than translation. Before one compares, one must translate. Only after translation comes the scheme of individual, species, and genus, of nationality and internationality.

At this time, I do not know whether a focus on bordering has gathered momentum across different disciplines in the humanities. However, this much can be asserted: bordering and translation are both problematics projected by the same theoretical perspective. Just as bordering is not solely about the demarcation of land, translation is not merely about language (in the sense of *langue*).

Hence, this chapter pursues a preliminary investigation into the discussion of translation, beyond the conventional domain of the linguistic. Yet the first issue that must be tackled is how to comprehend language from the viewpoint of translation, or, in other words, how to reverse the conventional comprehension of translation that depends on the trope of translation as a bridging or communication between two separate languages or *langues*.[9] But precisely because my approach is of a discursive analysis not confined to the domain of the linguistic or that is conducted irrespective of the rules, protocols, and assumptions of linguistics in general, it involves questions of figuration, schematism, mapping, cartographic representation, and the postulation of strategic positions, among others. In the conventional understanding of translation—elsewhere I have characterized it as "the schematism of co-figuration" (Sakai 1997: 1–17, 41–71)—the separation of two languages or the border between them is already presupposed. This view of translation always presumes the unity of one language (*langue*) and that of another, since their separation is taken for granted or already given; the contour of language is never understood to be something drawn or inscribed since it is supposed to be given in language itself. In other words, the conventional view of translation does not admit *bordering*.

Translation almost always involves a different language or at least a difference in or of language. But we must ask the same question again. What difference or differentiation is at issue here? How does it demand that we broaden our comprehension of translation? From the beginning, we have to guard against the static view of translation in which difference is substantialized; we should not yield to the reification of translation, for this would deny translation its potentiality to deterritorialize. Therefore, it is important to introduce the difference in and of language so as to comprehend translation, not in terms of the communication model of equivalence and exchange but rather as a form of political labor to create a continuity at the elusive point of discontinuity in the social. Above all else, translation is an act of sociality whereby the very relation between the addresser and the addressee is created, redefined, or modified. Then, what view of comparison can we acquire?

How do we recognize the identity of each language, or, how do we justify the presumption that languages can be categorized in terms of one and many? Is language a countable among the nominals, like an apple or an orange and unlike water? Is it not possible to think of language, for example, in terms of those grammars in which the distinction of the singular and the plural is irrelevant?

As I have repeatedly drawn attention to, what is at stake here is the unity of language, a certain *"positivity of discourse"* or "historical *a priori*" (Foucault 1972), in terms of which we understand a different language or difference in language. To put it slightly differently, it is to understand how we allow ourselves to tell one language from another, to represent language as if it were a unity.

My answer to this question, which I posed some twenty years ago, is that the unity of language is like Kant's *regulative idea* (Sakai 1992: 326). It organizes knowledge but is not empirically verifiable. The regulative idea does not concern itself with the possibility of experience; it does not partake of the constitutive faculty; it is no more than a rule by which a search in the series of empirical data is prescribed. It does not guarantee an empirically verifiable truth but, on the contrary, it forbids "[the search for truth] to bring it[-self] to a close by treating anything at which it may arrive as absolutely unconditioned" (Kant 1929: 450 [A 509; B 537]). Therefore, the regulative idea gives only an *object in idea*; it only means "a *schema* for which no object, not even a hypothetical one, is directly given" (550 [A 670; B 698]; emphasis added). The unity of language cannot be given in experience because it is nothing but a regulative idea that enables us to comprehend related data about languages "in an indirect manner, in their systematic unity, by means of their relation to this idea" (Kant 129: 550 [A 670; B 698]). This is to say that a language is an individual—an indivisible unity—only insofar as it is an object in idea, a regulative idea. It is not possible to know empirically whether a particular language exists as a unified individual or not. But by subscribing to the idea of the unity of language, we can organize knowledge about languages in a systematic and scientific manner.

To the extent that the unity of national language ultimately serves as a *schema* for nationality and offers a sense of national integration, the idea of the unity of language opens up a discourse not only on the naturalized origin of an ethnic community but also on the entire imaginary associated with national language and culture. Kant also qualifies the regulative idea as a schema, that is, an image, design, outline, or figure, not exclusively in the order of the idea, but also in the order of the sensory. It is perhaps necessary to assert again that the unity of national language is never given at the level of primordial receptivity but rather at the level of representation. In other words, the unity of language belongs to the dimension of a screen or a consciousness on which images are projected by our mind. In short, a language is an individual only as a schema. Therefore, this unity is in the order of image, figure, or schema: it comes into being in representation. And this figurative representation of national language allows us to discuss the communal experience of the common national or ethnic language. A language may be pure, authentic, hybridized, polluted, or corrupt, yet regardless of a particular assessment of it, the very possibility of praising, authenticating, complaining about, or deploring it is offered by the unity of that language as a regulative idea.

However, we all know that the institution of the nation-state is a relatively recent invention. It assumes the presence of a territorial national state sovereignty in the modern international world, so that it could not have existed before the internationality of the modern world was introduced. Thus we are led to suspect that the idea of the unity of language as the schema for ethnic and national communality must also be a recent invention, with a marked historicity.

Similarity beyond the Schematism of Cofiguration

Of course, *translation* is a term with much broader connotations than the transferance of meaning from one national or ethnic language into another, but in this chapter I am specifically concerned with the delimitation of translation, according to the *modern regime of translation* by which the idea of the national language is put into practice. I suggest that in this regime of translation, it is represented through a *schema of cofiguration*: only when translation is rendered representable, thanks to the schematism of cofiguration, does the putative unity of a national language ensue as a regulative idea.

This projection of schemata allows us to *imagine* or *represent* what goes on in translation, to give ourselves an image or representation *of* translation. Once imagined, translation is no longer a movement in potentiality. Its image or representation always postulates or projects two figures necessarily accompanied by a spatial division in terms of border. Insofar as the representation or image of translation, rather than the act of representation, is concerned, we are already implicated in the *tropes* and images specific to the modern regime of translation. As long as we represent translation to ourselves according to this regime, it is impossible to evade the *tropics of translation* particular to it. Primarily, border is a matter of tropics as far as translation is concerned because the unity of a national or ethnic language as a schema is already accompanied by another for the unity of a different language; this unity of a language is possible only in the element of many in one; and for there to be many, one unity must be distinguishable from another. In the representation of translation, therefore, one language has to be clearly and visibly distinguished from another. Consequently, the unity of one requires the postulation of a border or gap that separates it from another in the tropics of translation.

Translation takes various processes and forms, to the extent that it is a political labor to overcome points of incommensurability. It is nothing but a testimony to the universality of sociability. It need not be confined to the modern regime of translation; it may well lie outside this particular regime to which the majority of us are accustomed today. As a matter of fact, there used to be many different regimes of translation, but one by one such regimes were subdued to reach the stage where we are unable to apprehend translation in other terms than the schematism of co-figuration.[10]

Let us keep in mind, therefore, that in the modern international world, the unity of a language is always represented in relation to another unity; it is never given in and of itself, but rather in relation to an other. This is to say that the unity of language is possible only in the space of comparison. Internationality is implicit in the very notion of *a* language (*langue*), already when we represent to ourselves the whole or totality of a language as a unity.

One can hardly evade dialogic duality when determining the unity of a language; language as a unity almost always conjures up the copresence of another language. Yet, I cannot stress this point too strongly: the locale of comparison can never be identified cartographically with a national border on the geographic surface of the earth. The representation of translation in terms of the trope of border is nothing but an effect of the tropics of translation, precisely because it is preliminarily an act of drawing a border, of *bordering*, before it can be represented or imagined as a crossing of a border or a bridging of the gap allegedly existent between languages. The act of translation occurs in the place prior to the location where a border is drawn, in a *place before the area*. Thus, the locale of translation designates a place prior to the one cartographically assigned to it within the international world. It is in this sense that the locale of translation is *dislocated*, outside the system of geographic locations in the international world; it is also in this sense that the area is not only a determinate location within the striated space of comparison but also a performativity, a conduct whereby to introduce comparability.

One possible consequence we can draw from this discussion is that the locale of translation is potentially able to dislocate the system of allocation by which a language is located in the economy of nationality and internationality. In other words, to dislocate the putative coherence of the modern regime of translation. The locale of translation opens up the place of comparison, but it cannot be located within the configuration already prescribed by the schemata of individual, species, and genus. It is not located within the international world. On the contrary, it makes the location possible—in the sense of identifying an object within the already existing coordinate grids—of nationality and internationality; it facilitates the place of comparison, while pointing to a place without location.

By transnationality, I want to designate not the systematic of location configured by the logical economy of individual, species, and genus but rather the locale of translation that opens up the place of comparison. While internationality operates within the conceptual economy of the classical logic, transnationality undermines and reconfigures the schemata of nationality and internationality. It is in this sense that translation deterritorializes. And this deterritorializing potential of translation has been compulsorily reterritorialized by the schematism of cofiguration. Hence, transnationality indicates to us the locus of the foreign, something irreducible to the logical economy of individual, species, and genus. It may sound paradoxical, but translation takes place at the locale of dislocation; it dislocates the very distinction of inside and outside. It is the foreign that does not necessarily come from the outside of the national border; it is the foreign inherent within us all, regardless of whether within or outside of the nation. The conceptual topos of the foreign, where translation is in demand, can be found not in internationality but rather in transnationality, precisely because translation is prior to the determination of species difference.

Finally, we can return to the initial discussion of cultural difference once again. Now it is understandable why it is of decisive importance to discern two distinct approaches to cultural difference. The conventional notion of cultural difference, according to which the experience of discontinuity and nonsense is appropriated in the scheme of internationality, projects into our experience of cultural difference the very logic of individual, species, and genus and postulates the presence of the foreigner, foreign language, and foreign culture as an outside of nationality, spatially external to what John Stuart Mill called "a society of sympathy" (Mill, 1972: 391-409) outside of a somewhat essentialized realm of homogeneous national culture and language. But in our experience of cultural difference, the logic of inside as opposed to outside, which always determines the foreign as an intrusion of the outside, is not imminent. Cultural difference need not be regulated by the conceptual economy of the classical logic. For before the translation, cultural difference cannot be reduced to the species difference between one species and another in the generality of genus. Too often multiculturalism has been captured by the rhetoric of internationality and the logic of specific difference.

In this article I have tried to examine the notion of the area in the disciplinary formation of area studies. As we explore the area as a performativity, as a conduct, of establishing the order of comparison, we cannot overlook the affinity of the area to that of translation. Furthermore, the area of performativity that introduces the space of comparability is couched in the discourse of the-West-and-the-Rest, it serves to reinforce the bifurcation of the West and the Rest. Our attempt to delineate an economy of similarity and dissimilarity aims at the project of comparative humanities from the dominant mode of comparative nationality in such a way as to allow for configurations of similarities outside the economy of the internationality of race, ethnicity, culture, civilization, and nationality. Yet this effort is far from exhaustive: here I can only intimate how we can possibly conceive of knowledge production in area studies in a different direction, with a different thrust from the schematism of cofiguration.

Notes

1 Gilles Delueze and Félix Guattari introduced this term, *machinic phylum*, as a corrective of the *hylomorphic* model, the schema of matter and form. "The *machinic phylum* is materiality, natural or artificial, and both simultaneously; it is matter in movement, in flux, in variation, matter as a conveyor of singularities and traits of expression. This has obvious consequences: namely, this matter-flow can only be *followed*" (Deleuze and Guattari 1987: 409; italics in the original). What is of decisive importance for us is that the *machinic phylum* is something to be followed but not to be compared. In order to compare, a certain conduct is necessary.

2 Various historicist readings of Marx's (1990) *Capital* have obscured this point. The primitive accumulation of capital does not refer to a historical stage after which capitalism is established once and for all. What Marx called "the original sin of capitalism" must be repeated, and it is in this sense that the area is necessarily repeatedly committed.

3 It is important to note that the terms *system* and *systematicity* must be clearly differentiated.

A system is a series of differences most typically observable in the concept of paradigm. The identity of one term is always dependent on other terms in the system, but there is no assumption that this system is finite or circulatory, so that the system cannot be treated as if it were an enclosure or substance. Conversely, a systematicity is a set of logical relations that forms a totality and logical enclosure. In Saussurean linguistics, the concept of *language* is comprehended in terms of systems but never in terms of systematicity whereas the concept of *langue* assumes the grasp of languages as systematicities. To regard an ethnic/national language as an individual is to posit a language as a set of systematicities. What is conventionally referred to as grammar is the characterization of a language in terms of systematicity. In my examples, it is impossible to find any sensible ground that what is signified by "the Chinese language" can constitute an organic unity. One can only claim that it connotes an ensemble of languages that are used in the People's Republic of China or spoken by the people of Chinese ethnicity. Does it connote a group of languages that are affiliated with one another through some common phonetic, syntactical, or semantic features? Or does it imply the set of grammatical rules that the state of the People's Republic of China imposes on its population as the standard language? It is doubtful that the individuality of the Chinese language constitutes the priority of the empirical with which the individual has traditionally been endowed.

4   In regard to the concepts of being-in-common, communication and communion, I owe much to Jean-Luc Nancy's essay "La Communauté désoeuvrée" in *Aléa* (1983: 11–49). For the English translation, see *The Inoperative Community* (Nancy 1991: 1–42). Also important is Nancy's essay "De l'êÊtre-en-commun" in *La communauté désoeuvrée* (2004).

5   I learned the term *bordering* from Sandro Mezzadra and Brett Neilson's "Border as Method; or, The Multiplication of Labor." This is a paper that was presented at the International Conference "Italian as Second Language: Citizenship, Language, and Translation" in Rimini, Italy, on February, 2008. It is now available online in *transversal*, the journal of the European Institute for Progressive Cultural Policies (2008b).

6   The problem of *discontinuity* must be highlighted precisely because politics is a set of actions by which to create *continuity*. It is important to stress that to divide is not to introduce discontinuity. On the contrary, it is possible to divide only when the space in which a divide is introduced is already continuous. The presence of a border is a sign of continuity rather than discontinuity.

7   Historically, this thought habit owes much to what Carl Schmitt called the *Jus Publicum Europaeum* or the system of Eurocentric International Law. The international world initially outlined in the Treaty of Westphalia is conceptually dependent on the concept of the territorial state sovereignty, according to Schmitt. I follow his argument as far as the historical formation of the modern international world is concerned. See Schmitt [1950] 2006.

8   The term *national state* is not without problems. In the seventeenth century, when the Treaty of Westphalia was first introduced, there was no state that derived its legitimacy

from the new and peculiar communality called "the nation." The national state therefore does not mean the sovereign nation-state, which was not actualized until the eighteenth century.

9 A luminous examination of the notion of "communication" was carried out by Briankle G. Chang in his *Deconstructing Communication: Representation, Subject, and Economies of Exchange* (1996), from which I have learned much.

10 I discussed the history of a discursive formation in which the modern regime of translation replaced other regimes in Sakai 1992.

References

Baldwin, James. 1963. *The Fire Next Time*. New York: The Dial Press.

Chang, Briankle G. 1996. *Deconstructing Communication: Representation, Subject, and Economies of Exchange*. Minneapolis: University of Minnesota Press.

Deleuze, Gilles, and Felix Guattari. 1987. *A Thousand Plateaus: Capitalism and Schizophrenia*, translated by Brian Massumi. Minneapolis: University of Minnesota Press.

Foucault, Michel. [1969] 1972. The Archaeology of Knowledge and the Discourse on Language, translated by A. M. Sheridan Smith. New York: Harper & Row.

Hall, Stuart. 1996. "The West and the Rest: Discourse and Power." In *Modernity: An Introduction to Modern Societies*, edited by Stuart Hall, David Held, Don Hubert, and Kenneth Thompson, 184–227. Malden, MA: Blackwell.

Kant, Immanuel. 1929. *Critique of Pure Reason*, translated by Norman Kemp Smith. New York: St. Martin's Press.

Marx, Karl. 1990. *Capital*. Vol. 1. New York: Penguin Classics.

Mezzadra, Sandro, and Brett Neilson. 2008a. "Border as Method; or, The Multiplication of Labor." Paper presented at the International Conference "Italian as Second Language: Citizenship, Language, and Translation," Rimini, Italy, February 4.

Mezzadra, Sandro, and Brett Neilson. 2008b. "Border as Method; or, The Multiplication of Labor." *transversal*, March. www.eipcp.net/transversal/0608/mezzadraneilson/en (accessed February 8, 2017).

Mezzadra, Sandro, and Naoki Sakai. 2014. "Introduction." Special issue, *Translation: A Transdisciplinary Journal* 4: 9–29.

Mill, John Stuart. [1861] 1972. Utilitarianism. Liberty. Considerations on Representative Government. London: Dent.

Nancy, Jean-Luc. 1983. "La communauté désoeuvrée." *Aléa* 4: 11–49.

Nancy, Jean-Luc. 1991. *The Inoperative Community*, edited by Peter Connor, translated by Peter Connor, Lisa Garbus, Michael Holland, and Simona Sawhney. Minneapolis: University of Minnesota Press, 1991.

Nancy, Jean-Luc. 2004. "De l'être-en-commun." In *La communauté désoeuvrée*, 199–234. Paris: Christian Bourgois.

Sakai, Naoki. 1992. *Voices of the Past: The Status of Language in Eighteenth Century Japanese Discourse*. Ithaca, NY: Cornell University Press.

Sakai, Naoki. 1997. *Translation and Subjectivity: On "Japan" and Cultural Nationalism*. Minneapolis: University of Minnesota Press.

Sakai, Naoki. 2012. "*Positions* and Positionalities: After Two Decades." *positions: asia critique* 20, no. 1: 67–94.

Schmitt, Carl. [1950] 2006. *The* Nomos *of the Earth in the International Law of the* Jus Publicum Europæum, translated by G. L. Ulmen. New York: Telos Press Publishing.

Solomon, Jon. 2014. "Invoking the West, and Giorgio Agamben's 'Romantic Ideology' and the Civilizational Transference." *Concentric: Literary and Cultural Studies* 40, no.2. 125–47.

Winichakul, Thongchai. 1994. *Siam Mapped: A History of the Geo-Body of a Nation*. Honolulu: University of Hawai'i Press.

Wittgenstein, Ludwig. [1953] 1968. *Philosophical Investigations*, translated by G. E. M. Anscombe. Oxford: Basil Blackwell.

Contributors

Étienne Balibar was born in 1942. He graduated at the Sorbonne in Paris and later took his PhD from the University of Nijmegen (Netherlands). He is now emeritus professor of philosophy at the Paris Nanterre University and anniversary chair of contemporary European philosophy at Kingston University, London. He is coauthor or author of *Reading Capital* (1965, with Louis Althusser), *Race, Nation, Class: Ambiguous Identities* (1991, with Immanuel Wallerstein), *Masses, Classes, Ideas* (1994), *The Philosophy of Marx* (1995), *Spinoza and Politics* (1998), *Politics and the Other Scene* (2002), *We, the People of Europe? Reflections on Transnational Citizenship* (2004), and *Identity and Difference: John Locke and the Invention of Consciousness* (2013).

Ken C. Kawashima is associate professor in the Department of East Asian Studies at at the University of Toronto. He is author of *The Proletarian Gamble: Korean Workers in Interwar Japan* (2009), coeditor of *Tosaka Jun: A Critical Reader* (2013), and translator of Uno Kozo's *Theory of Crisis* (forthcoming from the Historical Materialism series at Brill Publishers). He is currently working on a book with Gavin Walker titled *Surplus alongside Excess.*

positions 27:1   DOI 10.1215/10679847-7251969

Sandro Mezzadra teaches political theory at the University of Bologna and is adjunct research fellow at the Institute for Culture and Society of Western Sydney University. He has been visiting professor or research fellow at several institutions, including the New School for Social Research (New York), Humboldt-Universität (Berlin); Duke University (Durham, NC), Fondation Maison des sciences de l'homme (Paris); University of Ljubljana, FLACSO Ecuador (Quito), and UNSAM (Buenos Aires).

In the last decade his work has particularly centered on the relations between globalization, migration, and political processes and on contemporary capitalism as well as on postcolonial theory and criticism. He is one of the founders of the website Euronomade (www .euronomade.info).

His most recent book is *Nei cantieri marxiani. Il soggetto e la sua produzione* (*In the Marxian Workshops. The Subject and its Production*, 2014, forthcoming in English). With Brett Neilson he is the author of *Border as Method; or, The Multiplication of Labor* (2013). Mezzadra and Nielson's new book, *The Politics of Operations: Excavating Contemporary Capitalism*, is forthcoming.

Tessa Morris-Suzuki is emeritus professor of Japanese history and Australian Research Council Laureate fellow at the Australian National University. Her research has focused on modern Japanese and East Asian history, including topics of memory and reconciliation, migration, ethnic minorities, and grassroots movements. She is author of *Borderline Japan: Foreigners and Frontier Controls in the Postwar Era* (2010), coauthor of *East Asia beyond the History Wars: Confronting the Ghosts of War* (2013), coeditor of *New Worlds from Below: Informal Life Politics and Grassroots Action in Twenty-First Century Northeast Asia* (2017) and *The Living Politics of Self-Help Movements in East Asia* (2018), and editor of *The Korean War in Asia: A Hidden History* (2018).

Naoki Sakai is Goldwin Smith professor of comparative literature and Asian studies at Cornell University. He has published in the fields of comparative literature, intellectual history, translation studies, the studies of racism and nationalism, and the histories of textuality. His publications include *Translation and Subjectivity* (1997), *Voices of the Past* (1991). *The Stillbirth of the Japanese as a Language and as an Ethnos* (1995), *Nationalism of* Hikikomori (2017), and *The End of Pax Americana and the Nationalism of* Hikikomori (forthcoming). He is editor of a number of volumes, including coeditor of *Politics of Translation*, special issue of *Translation*; (with Sandro Mezzadra, 2014) and *The Trans-Pacific Imagination* (with Hyon Joo Yoo, 2012). Sakai served as the founding editor for the project *TRACES*, a multilingual series in Korean, Chinese, English, Spanish, and Japanese.

Shu-mei Shih is professor of Asian languages and cultures, comparative literature, and Asian American studies at UCLA, with a fractional appointment as the Hon-yin and Suet-fong Chan professor of Chinese at the University of Hong Kong, and an honorary chair professor at the National Taiwan Normal University. Among other works, her book *Visuality and*

Identity: Sinophone Articulations across the Pacific (2007) has been attributed as having inaugurated a new field of study called Sinophone studies, and its Mandarin Chinese translation has gone into three printings (2013, 2015, 2018). *Sinophone Studies: A Critical Reader* (2013) is a textbook that she coedited for the field. Her latest work in this field is *Against Diaspora: Discourses on Sinophone Studies*, a monograph published in Taiwan (2017), which went into second printing in less than six months. She is currently working on her next book in Sinophone studies entitled *Empires of the Sinophone* as well as a book on comparative methodology entitled *Comparison as Relation*.

Jon Solomon is a professor in the Department of Chinese Literature, University of Lyon (Jean Moulin Lyon 3). His publications have focused on the biopolitics of translation, developing a critique of the disciplinary divisions of the humanities in their relation to the various economic and political divisions of the postcolonial world. Recent publications include an article in Chinese for the Taiwan journal *Router* about the relation between translation, logistics, and sovereignty in the context of the 2014 Sunflower Movement in Taiwan and an article in English titled *Logistical Species and Translational Process* that appeared in the Montreal-based journal *Intermédialités*.

Tazaki Hideaki is a professor in the Graduate School of Contemporary Psychology at Rikkyo University in Tokyo. He is the author of numerous works in contemporary theory, including *Uru shintai, kau shintai: Sekkusuwaaku-ron no shatei* (*The Selling Body, the Buying Body: On Theories of Sex Work*, 2015), *Yume no rōdō/Rōdō no yume: Furansu shoki shakaishugi no keiken* (*The Labor of Dreams/The Dreams of Labor: The Experience of Early French Socialism*, 2015), and *Munō na mono tachi no kyōdōtai* (*The Community of the Incompetent*, 2008), among others.

Gavin Walker is associate professor of history at McGill University in Montreal, Québec, where he works at the intersection of critical theory, intellectual history, and literary studies. He is the author of *The Sublime Perversion of Capital* (2016), the editor of *The Japanese '68: Theory, Politics, Aesthetics* (2018), and editor and translator of Kojin Karatani's *Marx: Towards the Centre of Possibility* (forthcoming). He is the author of over fifty articles and chapters on topics in modern Japanese and French thought, the history of Marxist theory, postcolonial studies, and intellectual history. He is currently finishing a new book, *Topologies of the Dialectic: Cultural Forms and the Allegories of History*. With Ken Kawashima, he is working on a new project titled *Surplus alongside Excess*.

Keep up to date on new scholarship

Issue alerts are a great way to stay current on all the cutting-edge scholarship from your favorite Duke University Press journals. This free service delivers tables of contents directly to your inbox, informing you of the latest groundbreaking work as soon as it is published.

To sign up for issue alerts:

1. Visit **dukeu.press/register** and register for an account. You do not need to provide a customer number.

2. After registering, visit **dukeu.press/alerts**.

3. Go to "Latest Issue Alerts" and click on "Add Alerts."

4. Select as many publications as you would like from the pop-up window and click "Add Alerts."

read.dukeupress.edu/journals    DUKE UNIVERSITY PRESS